Praise for
A Hodgepodge of Microbes

"Lucas Yang's *A Hodgepodge of Microbes* is a delightful gateway to the enchanting world of the microbes that surround us. With clear explanations and vivid images, the book leads young readers to a better understanding of microbiology and the role microbes play in our lives and well-being. It offers valuable insights for budding scientists and more broadly every young person interested in enjoying a healthy life."

– Janet Hindler, MCLS, MT (ASCP), F(AAM), microbiologist

"This charming and delightful compendium of microbes transforms the invisible world into an enchanting narrative filled with wit and charm. The author's infectious enthusiasm makes even the smallest organisms feel like fascinating characters in their own right. A truly captivating read that proves science writing can be both educational and utterly delightful."

– Grace Aldrovandi, MD CM, pediatric infectious disease physician and scientist

"Lucas Yang delightfully illuminates the microcosmos for readers of all ages, bringing whimsy and fun to sound science so we can better understand the microorganisms that share our planet (and our bodies). *A Hodgepodge of Microbes* is sure to please anyone curious about the little buggers that play such an integral, if unseen, role in our lives."

– Sandy Cohen, MPH, science journalist

A HODGEPODGE OF
MICROBES

A JOURNEY INTO THE MARVELOUS WORLD OF MICROORGANISMS

LUCAS YANG

A Hodgepodge of Microbes
A Journey into the Marvelous World of Microorganisms
by Lucas Yang

1. SCI045000 SCIENCE / Life Sciences / Microbiology
2. JNF051050 JUVENILE NONFICTION / Science & Nature / Biology
3. YAN050030 YOUNG ADULT NONFICTION / Science & Nature / Biology

ISBN: (paperback): 979-8-88636-053-0
ISBN: (hardcover): 979-8-88636-054-7

Library of Congress Control Number: 2025911692

Cover design by Lucas Yang and Lewis Agrell

Printed in the United States of America

Authority Publishing
13389 Folsom Blvd #300-256
Folsom, CA 95630
800-877-1097
www.AuthorityPublishing.com

To my family and teachers—and the microbes around us too.

CONTENTS

CHAPTER 1

MEET THE MICROBES

HEY READER. WHETHER YOU ARE A HUMAN (*HOMO SAPIENS*), A chimpanzee (*Pan troglodytes*), a velociraptor (*Velociraptor mongoliensis*), or a toadstool (*Amanita muscaria*), don't be afraid of the chonkiness of this book. This book is free of the pure boredom oozed out by dusty research papers or the humdrum reeking from moldering archives. You will be entertained with a concoction of wit-packed writing mixed with plenty of—gasp!—cartoons. This isn't AP biology after all!

Many of the microbes showcased in this book are harmful. This is a book about infectious diseases, after all. However, most microbes are tame, jelly-like creatures unseen by human eyes. Though we can't see microbes, the very backbone of life is built around 'em. They help clean up oil spills, manufacture detergent, and make greener alternatives to gasoline. A few microbes even rustle up some of your favorite foods, like soy sauce, yogurt, and cheese. And perhaps most importantly, microorganisms regulate the health of the soil, decompose dead stuff, and support the entire food chain. Without these little fellas, life on Earth would be impossible.

It's very important not to consider microbes as evil. These little beasties are an essential part of Mother Nature. They're just innocently trying to multiply and continue their species. Proof: There are more bacteria than human cells in your body. If you cringe

The labels in the illustration read: "Dileptus" and "Actinobolina".

whenever you hear the word "germ", this book will help you understand and allay your fears. Besides, the word "germ" isn't the proper scientific term for a microscopic, blobby organism. Try using the word "microbe" instead (quoting my fifth-grade science teacher). Now enough chatter. Let's get this microscopic party started!

Bacteria

Bacteria are some of the deadliest pathogens on this planet but also some of the most helpful: they are a vital part of our microbiome, a lively community of microbes that helps you digest food, absorb nutrients, and drive off invading pathogens. Bacteria are single-celled creatures that reproduce in a bonanza of different ways. Some of these critters divide in two (mitosis) or form tough, hardy spores (sporulation). Others grow buds, which detach as they mature, or split into multiple organisms when hacked apart. Bacteria multiply at a breakneck pace and reach staggeringly large populations: there are about five million trillion trillion individual bacteria on our planet. Bacteria come in a gamut of shapes and sizes, such as rods (bacilli) or spheres (cocci).

The cells in you, your pet goldfish, and the maple tree in your backyard are all eukaryotes. Eukaryotes are cells with a nucleus. The nucleus is a cell's home base: it shields their fragile DNA and tells the cell what to do, when to grow, and when to reproduce. On the other hand, bacteria are prokaryotes; they don't have a nucleus. Their genetic material just kind of drifts around inside them. Bacteria also have a freaky backup set of genetic material called a plasmid. Plasmids provide bacteria with nifty genetic upgrades that boost their chances of survival. They allow bacteria to spread faster or become resistant to antibiotics. These little tykes have tons of tricks up their sleeves!

Tropical King: *Thiomargarita magnifica*

These bacteria with a mouthful of a name are 0.79 inches (2 centimeters) long, boast half a million copies of DNA inside their cells, and are visible to the naked eye! (Usually, bacteria are only about 1 micrometer long, or 127,000 times smaller than your average hamster.) Microorganisms usually stay dinky because they absorb nutrients and oxygen through their cell wall in a process known as diffusion, which only works on a small scale. If they grew larger, most would suffocate. But this micro kaiju—which looks like a white strand of hair—has most of its cellular components wrapped in a thin layer around a gigantic microbial storage container known as a vacuole. This allows *T. magnifica* to grow outrageously tubby while still depending on diffusion. These roly-poly microbes are sulfur-fixing bacteria; to eat, they munch on nitrogen and elemental sulfur in the comfy sediment of mangrove swamps and stuff excess grub into their vacuoles for any lean times ahead. Strangely, *T. magnifica* cells share more similarities with plant or animal cells, not bacterial ones. Many questions still shroud these bacteria—they were just discovered in 2022—but one thing is certain: it's always good to have your own personal snack shack!

Tiny Titan: *Pelagibacter ubique*

Pelagibacter ubique is one of the smallest species of bacteria in the world. But this teensy tot packs a punch: it's the most common living thing on Earth. *P. ubique* happily floats in the sea, quietly multiplying and soaking up sunshine to make ATP—the energy currency used to power up all living things. However, unlike the oak tree in your backyard, *P. ubique* doesn't photosynthesize. The bacteria have a special, sunlight-activated protein called proteorhodopsin that uses the movement of hydrogen atoms to create ATP. *P. ubique* doesn't rely on the sun alone to get its grub either. The bacteria use special chemicals called ABC transporters that activate *P. ubique's* enzymes, which promptly gobble up the bacteria's chow. This complicated strategy is worth it when your stomach's rumbling at sea—*P. ubique* lives in areas that lack yummy nutrients. In conditions where other bacteria would starve, *P. ubique* doesn't bat an eye.

P. ubique's small size isn't a joke either. People believe *P. ubique* turned bite-sized to avoid bacteriophages, nasty viruses adapted for killing bacteria. A bacteriophage dubbed HTVC010P (scientists really should've chosen a catchier name) was caught red-handed infecting *P. ubique* in 2013, and there are probably more viral villains out there undiscovered pestering this tiny mighty. Who knew a bacteria so small could be so complex? So if you're picked on for being short, tell those big cheeses being small has its advantages!

The Astronaut: *Gloeocapsa magma*

No, these bacteria don't have anything to do with lava or any other molten rock that spurts out from a volcano. *G. magma* is actually a species of cyanobacteria—photosynthetic bacteria mistakenly called blue-green algae. These humble organisms often form disgusting mats of green sludge on ponds and lakes, but these unassuming critters are the reason you're alive today. Some 3.5 billion years ago, cyanobacteria gobbled up most of the carbon dioxide in Earth's icky, old atmosphere and released fresh, clean oxygen back into the air.

Besides sprucing up our atmosphere, *G. magma* has also gone to space: colonies of these invincible bacteria clung to the outside of the International Space Station for a year. Other microbes would've been fried to a crisp by cosmic radiation, but *G. magma* has the remarkable ability to stitch up its damaged DNA. These super-bacteria even watch each other's backs; when exposed to harsh conditions, they stick together in protective clusters. *G. magma* could save thousands of lives by teaching us how to protect ourselves from radioactivity, but this rascal mostly spends its time munching away at the limestone in roof tiles!

Virus

Viruses are tricky organisms. Even big-brain scientists squabble over whether these fellas are dead or alive. Viruses can't reproduce on their own. They have to infect a host cell to make baby viruses. Viruses aren't made out of cells either. If you're not made of cells, you're not alive. Because of this, some academics think viruses are just shells of molecules. Others object and say viruses are honest-to-goodness living things. In any case, these guys are a nuisance. Everyone agrees that viruses are just tiny, annoying protein bags constantly harassing organisms. If these rotters aren't making people feel miserable, they're skulking around in the background. However, humans can take advantage of the killing properties of viruses and let them work for us. Remember the bacteriophages we mentioned earlier? These biological missiles are experts at massacring harmful bacteria but are completely harmless to us. Someday when all antibiotics are rendered useless, bacteriophages will be our only hope!

Most Grandfatherly Virus: *Mimivirus*

Mimivirus is an enormous virus that hijacks amoebas to reproduce. Under a microscope, *Mimivirus* is so big and plump it looks like a bacterium. A related species dubbed "*Mamavirus*," may be even larger. However, these bumbling blobs of goo are more mysterious than at a first glance. It's possible a virus similar to *Mimivirus* gave rise to bacteria, which kickstarted the evolution of a bunch of hairless apes using computers from single-celled organisms. A team of scientists analyzed *Mimivirus's* genome and found it had genes that allowed it to convert proteins and sugar into energy—something most viruses can only dream of doing.

Unlike most viruses, which use molecules called RNA to carry their genetic info, *Mimivirus* is a DNA virus. RNA isn't exclusive to viruses; it carries out a variety of important functions in all other life forms. Because of this, it's possible to say we *all* originated from viruses! These new discoveries fuel the debate about whether viruses are just sandwiches of complex molecules or bona fide living organisms. So the next time you catch a cold, just consider the virus as a pesky distant cousin! Maybe pat it on its microscopic head so it'll leave earlier.

Good Viruses?

GB Virus C is a commensal virus that lives in harmony with us—it doesn't cause headaches or runny noses—that drifts around in the bloodstream. It may seem like an "Easter egg" lying around in the background, but this mellow fellow actually lowers the risk of other viral infections. Ebola—a virus feared for its abilities to cause heavy bleeding (hemorrhaging) and extensive organ damage—is partially held at bay by GB Virus C. HIV—another blood-borne virus that weakens the immune system—is also slowed down by GB Virus C. This defies the myth that viruses are just dirty duds... no, some of them can save lives. GB Virus C is even affectionately called the "Good Boy" Virus. But don't make up your mind just yet! GB Virus C is related to the Hepatitis C virus, which causes nasty liver infections. And some immunocompromised people infected with GB Virus C suffer from intestinal problems, bleeding, abdominal pain, and other complications. Looks like this "good boy" still has a mischievous streak!

Silent Killer: Polio

If you contracted polio, you would most likely feel...nothing. Most people infected with polio don't even realize they were infected. Five percent of victims experience flu-like symptoms and stiff limbs. The most serious form of polio—paralytic polio—is rare but devastating. Paralytic polio can ravage anybody, but most commonly affects toddlers under the age of three. The virus usually multiplies in the intestines but occasionally spreads to the nervous system. As polio furiously gnaws away at the nerves, patients suffer from intense pain, muscle spasms, and paralysis just a few days after infection. If the virus attacks the brain, victims struggle to breathe and swallow. Polio spreads via sneezing, coughing, contaminated surfaces, and even slinks into food and water. Tragically, most children were infected with polio before a vaccine was invented. And in the mid-1900s, polio claimed the lives of half a million people each year.

Fortunately, polio vaccines came to the rescue. Dr. Jonas Salk made a vaccine from inactivated viruses in 1955 and Dr. Albert Sabin developed an oral vaccine in 1961. After the WHO launched a polio eradication campaign, polio cases plummeted by 99 percent. Only seven countries today have any detectable level of polio transmission. Even though many viruses are heartless hoodlums, the story of polio shows how we can vanquish them if we put our heads together.

Archaea

Archaea look like bacteria, but they aren't bacteria at all. They are ancient life-forms that existed long before bacteria. Even their name means old; "archaea" is derived from the Greek word for old, "*archaíos*." Archaea also use different materials to make their cell walls. While bacteria use (tongue twister alert!) peptidoglycan and lipopolysaccharide, archaea use a more primitive molecule called pseudopeptidoglycan. Plus, archaea can only split in two or branch off to reproduce. Bacteria have *way* more choices when it comes to replication. While bacteria duke it out with viruses, archaea avoid them altogether. Most archaea are extremophiles, toughing it out in extreme environments such as hot springs and pools of acid to avoid viral attackers. However, some bolder archaea live in your mouth and digestive tract and thrive on methane. Let's take a look at some of these plucky prokaryotes below.

A Little Wee Fellow: *Nanoarchaeum equitans*

Measuring an utterly invisible 400 nanometers in length, an archaea species called *Nanoarchaeum equitans* is the smallest living thing on the planet. This little guy lives at the bottom of the sea but also languishes in Yellowstone National Park's hot springs. *N. equitans* might be a puny shortie in the wide world of archaea, but this bite-sized microbe has a best friend. *N. equitans's* bestie is a larger archaea named *Ignicoccus hospitalis*. *N. equitans* can't even survive without its BFF; it must be attached to *I. hospitals* to stay alive. Why? *N. equitans* is so, so, so, so tiny that it lacks metabolic genes, essential pieces of DNA needed to synthesize fats and the building blocks of proteins. To compensate, *N. equitans* steals tiny bits of nosh from its much larger friend. This parasitic archaea takes the phrase "best friends stick together" a little too literally!

Acidic Fanatic: *Picrophilus*

Picrophilus lives in brutally acidic environments. It frolics around in acid mine drainage systems, noxious sulfur springs, and undersea volcanic vents that could burn bone. *Picrophilus* can comfortably swim in sizzling acid due to its extraordinarily thick cell wall, which is reinforced with fatty substances called lipids. *Picrophilus* also has very small pores—tiny holes used to guzzle up nutrients—compared to other microbes to prevent acid from leaking in. Any harmful protons—the positively charged particles that make acid so dangerous—are quickly vacuumed up by a proton pump. So when life gives you lemons, count yourself lucky that you aren't melting in acid!

Beating the Heat: *Pyrococcus furiosus*

Pyrococcus furiosus means "rushing fireball." The name fits: this micro-hotshot not only looks like a blazing ball of fire but was first discovered in a hydrothermal vent near Vulcano Island, Italy. *P. furiosus* thrives when temperatures reach the boiling point—literally! To beat the heat, *P. furiosus* has proteins and enzymes made from a metal called tungsten. Tungsten is hard to come by at the surface, but the waters off hydrothermal vents are chock full of the stuff. While even some of the grittiest microbes would be barbecued if they got too close to a hydrothermal vent, *P. furiosus* relies on red-hot temperatures to survive: this archaea's molecules function best above 195 degrees Fahrenheit (90 degrees Celsius). Asides from having muscles—er, molecules—of steel, *P. furiosus* can also withstand radiation hundreds of times higher than levels emitted from a nuclear bomb. Iron Man would be proud.

Protozoa

Protozoa, confusingly, belong to the protist kingdom—a catch-all term used to classify organisms that aren't fungi, animals, bacteria, or plants. Protists are a motley bunch that ranges from slime molds to algae to seaweed. Protozoa themselves are just as zany: they're a smorgasbord of single-celled animals. Some protozoa are amoebas, squirming around as they shapeshift their cell walls. (Protozoa are often mistakenly called "amoebas" as a whole, but protozoa aren't all amoebas.) Others are flagellates, which tumble about whipping flagella—think microscopic tails. Still more are ciliates, which haul themselves along on tiny hairs (cilia). Protozoa come in a medley of different personalities, too. Some are the biology teacher's pet, while others might eat the biology teacher's brain. Literally! But protozoa all have two things in common: unlike bacteria, they have a nucleus that safely wraps up their DNA in a rubbery ball (remember the eukaryotes we talked about earlier?). Protozoa can also engulf other cells—like animals eating food—while bacteria can only absorb nutrients through their cell membranes.

Teacher's Pet: *Paramecium*

Paramecium is the quintessential ciliate. This slipper-shaped fella zips through the water at a blistering speed of twelve body lengths per second. If we scaled up a *Paramecium* and raced it against a cheetah, it would be a close contest: the speedy feline can manage sixteen body lengths per second. But speed isn't the only trick up *Paramecium*'s sleeve. While most *Paramecium* chomp on bacteria and yeast, others store algae inside their bodies. *Paramecium* gives its photosynthetic pals a home, but the algae produce a yummy sugar called maltose—commonly known as corn syrup. *Paramecium* also wolfs down harmful fungi such as *Cryptococcus*—a dangerous brute that infects the brain. In the classroom, teachers love peering at *Paramecium* under the microscope. In the lab, scientists conduct genetic research on this jolly ball of furry jelly. Alas, it's a protozoa eat protozoa world out there: a tiny ciliate called *Didinium* loves gorging on *Paramecium*. This puny rapscallion can eat *Paramecium* twelve times its size!

Brain Eater: *Naegleria fowleri*

Naegleria fowleri stalks puddles of warm water worldwide. It usually feasts on bacteria, but pulls a switcheroo and causes lethal brain infections when it gets inside a human. Unsurprisingly, the media dubs this tiny terror "brain-eating amoeba." *N. fowleri* infections begin when the protozoa sneak up the nose and into the noggin. (Surprisingly, if you swallowed water containing *N. fowleri*, you would be completely fine.) The immune system sends out swarms of white blood cells like neutrophils to stop the protozoa. Individually, neutrophils have no chance against the mighty amoeba, but the white cells will surround and dogpile the parasite to kill it. Sadly, some of these micro zombies can sneak past the defenders. The body's faithful watchdogs, sentinel cells, often fail to detect *N. fowleri* and the white blood cells never come. If the protozoa reach the brain, the body sends out a hurricane of inflammatory chemicals called cytokines and mobs of white blood cells. However, *N. fowleri* almost always wins the intense melee that follows. Symptoms are mild at first—stiff neck, vomiting, fever, headache—but rapidly progress to seizures, confusion, hallucinations, coma, and eventually, death.

Fortunately, *N. fowleri* infections are extremely rare. You have a 1 in 34 million chance of contracting the protozoa. For reference, you have a 1 in 170 million chance of dying from a coconut falling on your head. It's even possible most people have been exposed to this evil beastie, fight it off, and never develop an infection. For now, this creeping menace of the swamp stays hidden in the muck. But as temperatures skyrocket and waters warm due to climate change, there's no knowing when a zombie apocalypse will occur...

Fungi

When it comes to fungi, mushrooms hog the limelight. However, molds and yeast are also part of the fungi family. Fungi live behind cozy cell walls made of chitin—the same stuff that forms insect shells. Some fungi are unicellular and tough it out alone. Others are multicellular and buddy up to form complex organisms. Despite appearances, fungi are related to animals, not bacteria or plants. This means the drugs used to treat them can be toxic to us. Fortunately, not all fungi are micro meanies. Molds whip up blue cheese. Yeasts help bread rise and make wine. Some fungi are even used for pest control. We'll showcase some of these slimy souls here.

Miracle Drug: *Penicillium chrysogenum*

Penicillium chrysogenum is the lovable fungus used to make the antibiotic penicillin, which knocks out bacteria by shredding their cell walls. We take penicillin for granted, but *P. chrysogenum* uses the drug to attack hungry microbes competing with it for food. As bacteria and mold brawl, we get the better end of the deal: a nifty medicine.

Penicillin was discovered by Alexander Fleming in 1928. While vacating for a few days, some *P. chrysogenum* mold squirmed into one of his petri dishes. When he returned, his petri dishes were covered in bacteria—save for one spot where a blob of *Penicillium* mold grew. Fleming realized the mold killed the bacteria. The discovery seemed promising, but penicillin didn't catch on as an antibiotic—not at first. Going from mold mush to pure penicillin was a big step; the antibiotic was almost impossible to purify. Penicillin was only tested on people in 1941 when forty-three-year-old policeman Albert Alexander became the first recipient of the drug. Albert was suffering from a devastating wound infection he acquired from pruning roses. After receiving penicillin, Albert rapidly recovered...but there wasn't enough of the antibiotic. The bacteria made a comeback and Albert died. Fortunately, penicillin was given a second chance: the drug successfully cured a fifteen-year-old boy of a *Streptococcus* infection later that year.

Luckily, we don't need to worry about penicillin shortages as much anymore. We farm *P. chrysogenum* in huge factory tanks churned by fans. The molds thrive on sugars and other nutrients dumped into the bubbling, frothing tanks. Penicillin is then purified and ready to go. Thanks to Fleming's science experiment gone wrong, penicillin has saved more than 500 million lives. Three cheers for penicillin!

Sugar Beasts: Yeasts

Yeasts are adorable, orb-shaped fungi with a sweet tooth. Yeasts get energy by fermenting sugars into alcohol. This creates tons of carbon dioxide bubbles, which make bread fluffy and airy. Yeasts don't just chow down on bread; they also scoff on honey and grapes. Aside from giving bakers a hand in the kitchen, yeasts love helping other microbes too. In SCOBY (**S**ymbiotic **C**ulture **O**f **B**acteria and **Y**east) mixtures, yeasts and bacteria produce scrummy nutrients for each other. These SCOBYs help brew a delicious tea called kombucha.

However, yeast can be nasty, too. A yeast called *Candida* usually lives harmlessly on skin, kept in check by your microbiome. But when *Candida* levels careen out of control, the yeast infects the mouth and forms icky white lumps that look and feel like cottage cheese. This is called thrush. Thrush is usually a mild infection easily treated by antifungals. The real trouble begins if *Candida* causes invasive candidiasis. Invasive candidiasis occurs when *Candida* creeps into the bloodstream, bones, brain, and heart. Invasive candidiasis is one of the main causes of death in nursing homes; the weakened immune systems of the elderly provide a perfect opportunity for *Candida* to strike. Invasive candidiasis is so deadly that thirty percent of infected patients die. These villainous yeasts are also becoming superbugs: deadly microbes equipped with antifungal resistance. So while some yeasts are cutesy, sugar-loving gourmands, *Candida* is one yeast you definitely want to keep an eye on!

The End?

The wonderful world of microbes is in front of you, under you, and even on you. Microbes are glorious and whimsical organisms, ranging from tough cookies like *P. furiosus* to body battlers like GB Virus C. Though the majesty of big, brawny animals like lions, tigers, and elephants overshadow the microcosmos, sometimes it's nice to peer under a microscope and look through another world. There are millions of fascinating microbes out there waiting to be discovered, if only people would look at them. This book is just a humble depiction of this wondrous world, just one piece of the beautiful puzzle before your eyes. However, we hope this book sparks your curiosity about microbes and activates your inner microbiologist. Ready to enter a marvelous new world?

CHAPTER 2

LIVING WITH MICROBES

WHEN YOU WERE BORN, YOU WERE IMMEDIATELY EXPOSED TO all sorts of microbes in the environment. That basically means that you were literally skydiving into a pool of microbes. As unpleasant as this might sound, this is actually good for you. Your immune system needs to learn how not to fight off these itty-bitty creatures and just let them be. If your immune system lets beneficial microbes—so-called "normal flora"—stick around, they become part of your microbiome. Less than 0.0001 percent of the world's microbes wreak havoc. Scientists consider microbiomes to be a whole separate organ essential to human health as we know it. But not all microbes lend a hand. Some microbes hanging out in your body don't seem to do much of anything and just loaf around. Some lurk in the nooks and crannies of your body, waiting for you to become sick from separate causes before bursting out at the right moment to attack. And others are plain bad guys that try to make you miserable!

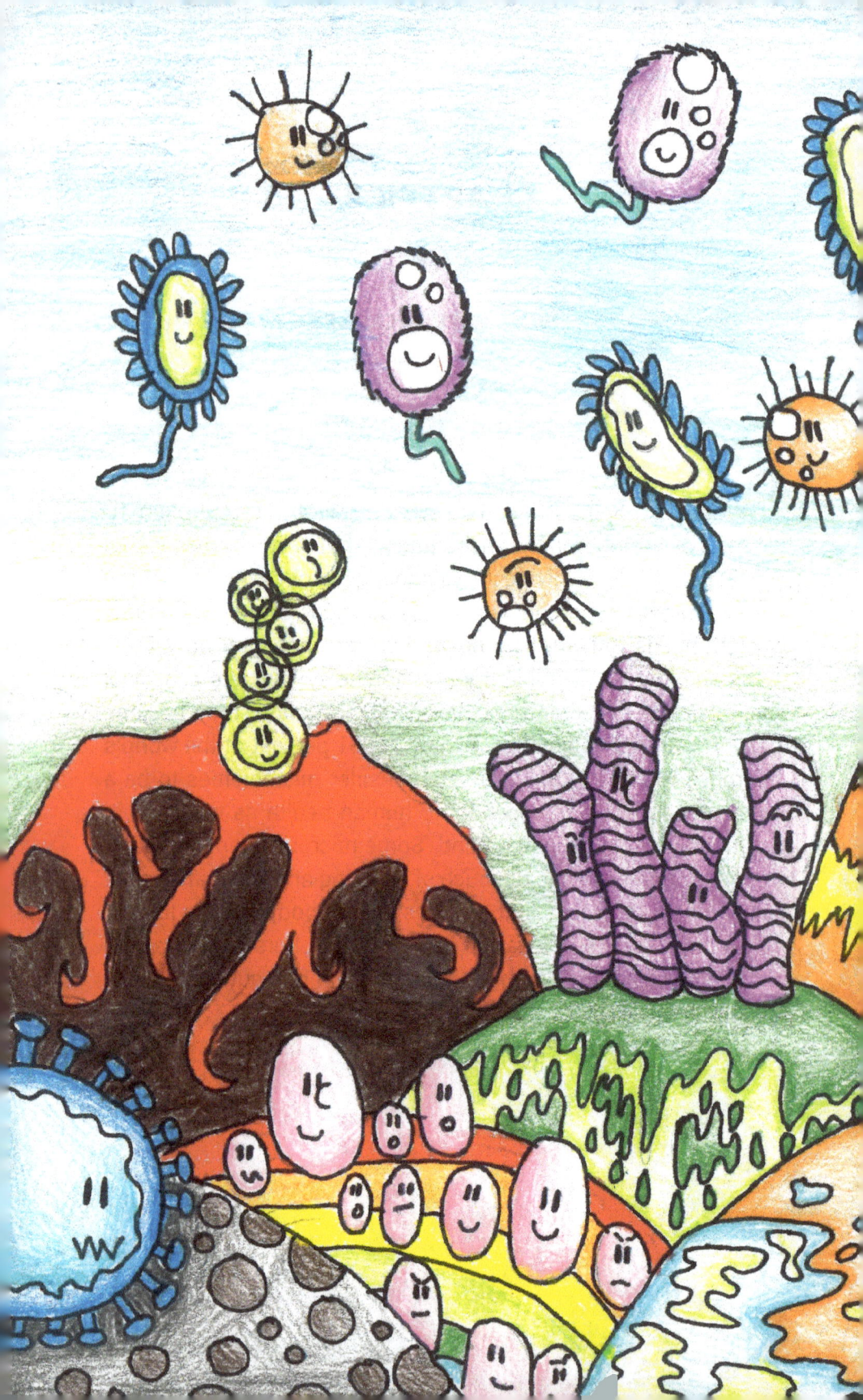

Your Marvelous Microbiome

Microbes in your microbiome act like servants, breaking down large molecules into absorbable nutrients, producing vitamins that the human body needs, crowding out harmful bacteria, and keeping the gut tightly sealed to prevent harmful compounds from leaking into the bloodstream. While the gut microbiome usually receives the most attention, the other microbiomes on and inside you are also important. A healthy skin microbiome is shown to decrease the risk of skin cancer!

Microbiomes are essential to animals, too. Old McDonald's cows couldn't digest their cud without the help of bacteria that bust cellulose in plant cell walls, nor could a gazelle on the African savanna or a panda in a bamboo forest. Anglerfish rely on bioluminescent microbes to lure in their prey. A species of wasp can't even reproduce without a strain of bacteria known as *Wolbachia*. Germ-free lab rats suffer a cocktail of nasty illnesses. Even caterpillars, which don't have a microbiome, still depend on wayward microorganisms to digest their grub (no pun intended).

The microbiome has become a hot topic in the field of microbiology lately, and recent studies have revealed a lot about your personal ecosystem. Like your fingerprint, no two microbiomes are alike; they are as different as the Sahara Desert is from the Amazon Rainforest. There's no one-size-fits-all microbiome. Microorganisms come and go. A healthy microbiome for a child might be entirely different for a grandpa. You receive about sixty percent of your microbiome from your mother during and after birth; the other forty percent is influenced by factors like disease, age, and diet. Exposure to the outside world also helps build up your microbiome. An intriguing study has revealed that the longer you spend time with someone, the more similar both of your microbiomes become—birds of a feather really do flock together.

Uninvited Guests

There are also microbes that just loaf around in the ol' human body hotel. A large group of archaea called methanogens love frolicking around in our digestive system but do no harm. The Torque Teno Virus (TTV) is another hitchhiker often found in the bloodstream. We don't know exactly what the virus does, but it doesn't seem to bother most folks. However, the jury is still out for TTV. Scientists are suspicious the fishy fellow causes graft—a fancy word for tissue—rejections during surgeries.

Of course, there are gangsters that try to take over the body too. However, other microbes and the immune system keep them at bay. Take the herpesviruses. There are eight types of these nasty buggers and I'll introduce some of 'em here.

Epstein-Barr virus (EBV)—increases cancer risk.

Cytomegalovirus (CMV)—attacks organs, especially those of transplant patients.

Varicella-Zoster Virus (VZV)—more commonly known as chickenpox—causes a painful rash.

Most humans are infected with these gnarly viruses during childhood, but the immune system fights them off. Unlike most infections, however, herpesviruses don't go away. They hide in the spinal cord or in lymphocytes, macrophages, and monocytes (specialized, jacked-up white blood cells) in the bone marrow. The immune system keeps a watchful eye on these gangsters, but in some unlucky folks, the body's police department gets out of whack and breaks down. This is called immunocompromising, which allows herpesviruses to burst out of their hiding place and attack the body...again. Since some of them hide out in neurons, these viruses can attack the brain and cause serious conditions such as meningoencephalitis—deadly brain inflammation. And when VZV strikes back, it causes an even more painful rash called shingles.

Enterococci bacteria are also among the gangsters, but only when they go bonkers. Normally, these bacteria help digest food and are valuable members of the gut microbiome. However, when a person is immunocompromised or has a leaky gut, enterococci march into the bloodstream and cause deadly sepsis. That's not just bad news—it's apocalyptic. Enterococci are resistant to a laundry list of antibiotics and are very difficult to shrug off.

The BK virus is usually a meek little ragamuffin that lives in the urinary system but flips out in immunocompromised patients. If left unchecked, the virus causes deadly bleeding in the bladder. This virus was *not* named after Burger King; instead, this rascal was named after the initials of a 1970s kidney transplant patient when it was first discovered. Are these microbes friend or foe? You make the call. But they live among us, and we don't have any say in the matter.

Spillover

Microbes have made themselves at home in our globetrotting industrial civilization. Unfortunately, speedy transit times and bustling worldwide trade mean harmful microbes in the environment can hitchhike on planes, boats, and trucks, and they spread to us in the blink of an eye. COVID-19 is certainly a well-known example of this, but these "spillover events" are nothing new. Lethal pathogens are often created when humans mess around with animals. An example is the lethal *Bacillus anthracis,* or anthrax. Anthrax is spread by airborne spores but also through wounds from infected horses, cows, or sheep. Fortunately, you only run the risk of catching this nasty bacterium in rural areas. Smallpox—one of the most dangerous viruses to human health—has been eradicated. However, its less-deadly predecessor, taterapox, still lives today, mainly infecting rodents. One of its cousins, monkeypox, began causing an epidemic in 2022, infecting tens of thousands across the globe. *Shigella dysenteriae*—a bacterium that causes bloody diarrhea—used to be very common before pasteurized milk came to the rescue.

Yersinia pestis, aka the Black Death, killed a third of the population in Medieval Europe. Despite all the carnage it caused, *Y. pestis* evolved from a wee lil bacterium—but only a few mutations altered it into a murderous microbe. A gene producing a protein called Pla was altered, allowing this microbe to infect the lungs. Another mutation helped *Y. pestis* replicate in fleas, which allowed it to spread like the plague (literally). This bacterial grim reaper originated from innocent-looking (and rather chubby) rodents called marmots in Kyrgyzstan. The first outbreak—confirmed with DNA sequencing from the teeth of ancient victims—occurred 1,300 years ago. HIV—believed to have emerged from chimps in the 1940s—hijacks the immune system by munching on white blood cells. It kills one million people a year and is the leading cause of death in some countries. Though drugs keep the virus at bay and significantly prolong lives, we're still living in the same HIV outbreak that's been raging since the 1980s.

Zoonotic, or animal-borne, diseases like these are the main challenges for healthcare. Getting rid of them isn't always possible: the flu, aka *Influenza*, spawns thousands and thousands of variants and infects multiple animals—from birds to pigs to humans—as hosts. Chickens and pigs are bred for meat around the globe (after all, when was the last time you gobbled down some chicken nuggets or a slice of bacon?), making flu transmission inevitable. Migrating geese spread the virus from place to place. The flu has eight segments of genetic info, but can swap segments with other variants, creating a Frankenstein of deadly, pandemic-causing strains. The most famous of these pandemics occurred in 1918 (read the next chapter for more info about this devastating outbreak). Most of our immune systems were hopelessly wet behind the ears and had no immunity to fight this brand-new virus.

You are What You Eat

Though it means grappling with pathogens, our cushy globalized civilization means we can walk into a supermarket and waddle out with our arms full of nutritious food and enjoy access to unprecedented advancements in healthcare. After all, our hunter-gatherer ancestors would have died to get their hands on fresh bell peppers or measles vaccines. However, smothering the world in metropolises also inflicts negative effects on our well-being. For instance, citizens dwelling in urban jungles have higher risks of developing obesity, inflammatory bowel disease, colon cancer, and heart disease. An ailing gut microbiome is suspected to be one of the factors responsible. Some people are quick to blame their microbes for acting up, but the old saying is still true: you are what you eat. Diners who pass the leafy greens and go carnivore unintentionally weed out the fiber-busting microbes, while nom-nomming on foods high in unhealthy fats and processed food can slash microbiome diversity. Loading our bodies with antibiotics also alters the composition of the microbiome. In-depth experiments on the gut microbiome include monitoring colon cell gene expression to see how the gut reacts to different microbial communities.

Where does the microbiome fit into all of this jumble of activity? Turns out your gut bacteria produce mood-influencing neurotransmitters like serotonin—commonly known as the "happy hormone"—that scurry up the vagus nerve and into your noggin. It goes both ways: the brain and gut affect the microbiome by altering its environment with things like diet choices. In short, by passing the sweetmeats and grabbing fiber-rich foods—such as avocados, bananas (personal favorite), almonds, or even air-popped popcorn—your microbes (and you) will jump for joy.

The Problem with Probiotics

Probiotics are dubbed "friendly bacteria." You've probably heard a lot about these fellas lately. Most people know probiotic bacteria hang out in yogurt and cheese, but they also happily settle down in other fermented foods like sauerkraut and kimchi. People suffering from stomach trouble rush to gulp down probiotic pills and drinks chock full of probiotics—or so it seems. The most concentrated probiotic sachets only contain 100 billion or so bacteria. This sounds like a lot of microbes, but the gut has 100 times more. The probiotic bacteria found in yogurt are also not key players in our guts. Most of them are only there because they manage to keep a toehold after being zapped by the pasteurization process, tumbled through a packaging plant, and scorched by stomach acid. Many small studies conducted on these "good bacteria" show promising results, but it's a tough job comparing the results due to inconsistencies. However, gobbling down these bacteria can't hurt: guzzling probiotic-rich foods can potentially shorten bouts of diarrhea. In premature infants, probiotics may also prevent the dreadful necrotic bowel disease. So far, the jury is still out on probiotics.

Get Dirty

According to a 2015 study by Norwegian scientist Ekaternia Avershina and her colleagues, vacuuming the house twice or more each week could potentially alter gut microbiome in pregnant women and young children. However, the study never discovered any negative health consequences from excess vacuuming, so it's not a great excuse for skipping your household chores! However, getting dirty is one way backed by science to show your microbiome some love. Smothered under asphalt, it's easy to forget how important soil is to us. Water runs through it, we walk on top of it, crops grow in it, creatures burrow through it, and billions of microbes thrive in it. Yet we're all so keen to get rid of it. Compared to rural folks where becoming a bit grimy is the norm, urban dwellers have less diverse microbiomes—which may increase the risk of contracting gastrointestinal diseases—and weaker immune systems. For most of us city slickers, letting nature into our homes is enough. Strolling in the park and gardening are all ways to positively influence microbiome health. Don't forget about pets too! Curling up with a cat or a dog can potentially strengthen your microbiome and decrease the chances of contracting nasty allergies. Even cracking open a window does a world of good. Hospitals with a healthy breeze indoors invite harmless microbes from the great outdoors into their facilities, crowding out harmful pathogens.

The evidence is clear. As long as your immune system is strong, playing in dirt is awesome. We were meant to muck around with our microbial buddies. Uninvited guests and spillover events are just how we roll. Instead of trying to force them out of our world, we need to learn to live with them. Microbes are inextricably tied to our lives. They aren't just passing by—they run every aspect of our body. These tiny mighties digest our food, train your immune system, and keep you lively and chipper. However, if they are neglected, microbes will go out of their way to cause disease. Even your resident microbes can go renegade and attack your own body. Small day-to-day choices (e.g., eating your greens, having an active lifestyle, staying up-to-date on vaccines, keeping your distance from wild animals, responsible use of antibiotics, and proper hygiene) add up and have tremendous effects on the microorganisms in and around you. Our world is full of microbes—good, bad, and somewhere in between—and there's no denying it. So eat your veggies and get some sun outdoors. Your microbes will be happy and your body will thank you.

CHAPTER 3

THE RNA WORLD

W HO WAS YOUR FIRST GRANDFATHER OR GRANDMOTHER? THE answer is extremely murky. With our current evidence, it seems to be a proto-virus. This little guy was almost like a complex molecule instead of a virus and relied on RNA (not the more famous DNA) to spring into life. These micro-critters lived in an ancient world very different from today. About 3.8 billion years ago, the atmosphere was made out of volcanic gasses such as methane, carbon dioxide, and nasty hydrogen sulfide. The ground was young. Rivers of lava crisscrossed the ground. Comets battered the earth, crashing into the ground and bursting apart. Earth was a hostile place. Welcome to the RNA world, the dawn of life!

Ironically, these comets deposited molecules essential to life. The ice from these comets helped form the oceans. They also released amino acids, the molecules that join together to form proteins. It was a step to forming the first life. You might've heard about proteins before. They're the stuff found in a filet mignon or a bowl of tofu, and nutritionists go bananas about getting enough of it. But what exactly is a protein? Well, proteins are the building blocks of life. They act like the steel or wood frames propping up a building. Proteins literally hold a roof over your head, as well as the ones your virus ancestors had. These molecules shielded life from extreme temperatures and attacks from other organisms. Protein chemical names are also some of the longest words on Earth. The

human protein titin, which helps power your muscles, has a chemical name that could fill up half of this book and takes two hours to pronounce! But no matter what the name or function, all proteins help build the body.

But they can't do it alone, and that's when lipids and sugars come into the picture. Never mind. They were already on the comets in the first place. Lipids, more affectionately known as fat, are what cell membranes are made of. These molecules store energy and ferry messages across the body, such as in the form of hormones. Sugars like glucose are the juice that keep cells running, but some, such as starch and glycogen, team up with lipids to stockpile energy in the long run. Others are in cahoots with proteins to provide structural support. A sugar called cellulose makes the cell walls of plants, while two other types of these molecules, deoxyribose and ribose, play a crucial role in creating DNA and RNA, respectively.

DNA? RNA? The Difference Matters

While proteins, sugars, and lipids help bind everything together, DNA and RNA are like instruction manuals. They tell your cells what to do. Both are dubbed nucleic acids. But unlike the bubbling, neon-green stuff in the movies, nucleic acids won't burn you. You may already be familiar with DNA. It looks like a twisted ladder, or double helix. However, this life-giving molecule is more complex than you think. The rungs of the ladder are a bunch of chemicals called nucleotide bases called adenine, guanine, thymine, and cytosine. In reality, DNA doesn't look smooth and curved; it looks like a molten mess. Despite its topsy-turvy appearance, DNA is extremely stable, which is a good thing for life on our planet!

Many believe James Watson and Francis Crick discovered DNA. French scientist Friedrich Miescher found it first. When researching nucleic acids, he found a new guy hanging out in the nucleus and called the material "nuclein." But regardless of who discovered it, DNA usually gets much more attention than RNA, even though RNA probably started life in the first place. (It's not fair!) Unlike DNA, RNA looks like a thorny vine up close, though some rare viruses have double-stranded RNA. RNA is one freaky fellow; it's made up of adenine, guanine, cytosine, and uracil instead of the more famous combo of adenine, guanine, cytosine, and thymine in DNA.

RNA gives viruses their genetic input. This makes a virus extraordinarily powerful. RNA is much more versatile than DNA due to its flexible shape and characteristics. These RNA are similar to proteins and can also function as enzymes to start chemical reactions. To sum it up, RNA is the only molecule that can be a genetic instruction manual and build other chemicals itself at the same time. It doesn't need to rely on other molecules to complete the job. An example is shown in the next paragraph. See you there!

The Three Musketeers

Being the great-great-great-great (times a billion) grandchildren of viruses, it's no surprise we have some leftover RNA in ourselves. RNA is not only useful; it is an evolutionary clue we came from viruses. We have three types of RNA that help produce proteins— the Three Musketeers of protein synthesis, if you will. Ribosomal RNA is like an assembly factory. This little fellow helps produce proteins from individual amino acids. Transfer RNA is like a truck driver, delivering specific amino acids to the hard-working ribosome. Messenger RNA is a mailman and carries instructions from DNA to its ribosome factories.

This process run by RNA is so essential that it hasn't changed much since the dawn of life. All organisms, from the teeniest protozoa to the largest blue whale, use RNA! Ribosomal RNA in particular can also be used by microbiologists to see if a species has evolved. If it has, its ribosomal RNA sequence will change and remain the same until that species evolves again. Many of you probably thought RNA was a primitive, obscure thingamabob exclusive to viruses. Now we prove you wrong! But no matter how primitive, almost everyone is afraid of RNA viruses.

Fighting Back

Humans load ourselves with vaccines in order to stay safe from viruses. Insects have a protein called defensin that protects them from pathogens. Plants sabotage viruses by slicing up their genetic info. Fungi produce chemicals that kill viruses. Bacteria evolved special methods to escape bacteriophages (read more about them in Chapter 19) by changing their receptors—microscopic keyholes used to receive signals from other bacteria and gobble down nutrients.

Archaea, cousins of eukaryotes that don't cause disease, on the other hand, get around this by fleeing to remote areas that are either boiling hot, freezing cold, jam-packed with nuclear radiation, saltier than your grandpa's hotdogs, acidic enough to melt iron, or under so much pressure most organisms explode. RNA molecules are so unstable in these conditions that they burst apart. Archaea almost never get sick unless they venture out of their protective havens...but would you trade places with them just to avoid the seasonal sniffles?

An Odd Bunch

All viruses are grouped into seven different classes. Like a motley bunch of middle school classes, all of them are quite...weird. Class 1 are double-stranded DNA viruses, such as bacteriophages. Class 2 are rare, single-stranded DNA viruses. One of them, called parvovirus, causes a cold-like illness and likes to infect your red blood cells. Infections can lead to anemia, or a low red blood cell count. This is bad; red blood cells get that good old oxygen to the rest of your cells. Without them, you're kaput. Class 3 are double-stranded RNA viruses, which include the rotaviruses.

These microbial crooks caused extreme stomach trouble before a vaccine was available. Class 4 are single-stranded, positive-sense RNA viruses. These include rhinovirus—or the common cold virus.

Class 5 are negative-sense RNA viruses. These viruses are truly scary and include the grim reaper of the viruses, Ebola, which causes people to bleed and vomit to death. Class 5 viruses must convert their negative-sense into positive-sense RNA first before they can settle down and make proteins. On the other hand, positive-sense RNA viruses can get to work immediately. Class 6 are a very strange group of viruses that are first RNA viruses, transform into DNA viruses, and go back to being RNA viruses. These are also known as retroviruses. A famous example is the once deadly HIV. Fortunately, HIV's white-blood-cell-killing tactics have been humbled by effective drugs, allowing infected people to live long and healthy lives. Class 7 consists of DNA viruses that transform into RNA viruses...then change into DNA viruses once more. They also use their shapeshifting abilities to reproduce. One of them is the nasty Hepatitis B virus (HBV), which causes severe liver problems and cancer.

But regardless of their differences, all viruses have a few things in common. One, they inject their genetic information to hijack a host cell so they can reproduce. Two, they cause the host to fight back and mutate to better fight the viruses, creating host genetic diversity. This virus-and-host interaction may have contributed to the development of advanced life forms. Battles between viruses and their hosts have been traced back as far as the RNA world almost four billion years ago. Speaking of which, I think it is a good time to talk about those old-timer viruses once more.

By Hook or by Crook

Early viruses already had the basic features of today's viruses. They were also all SELFISH. In a sense, viruses are greedy war-lords because they cause harm to the host for their own good.

But that's not the point. RNA is quite unstable, so viruses need to replicate quickly. There's no time for significant changes, so the same RNA sequence is used over and over and over again, "self-ishly" creating the same virus. This limits the gene pool of a viral population. Fortunately for them, mutations do happen during replication. Some mutations can be bad and cause harm.

However, a few lucky viruses develop helpful mutations that make them replicate faster or protect them from outside stress. These beneficial mutations accumulated and eventually led to the evolution of more complex life. If you are wondering what happened to the RNA viruses all those years ago when they had no non-viral hosts, it may not surprise you that viruses can infect other viruses! However, evolution doesn't hurry. It took a billion years for the soupy RNA world to become diverse enough to eventually form primitive cells, and another two billion years for those cells to clump together and form animals. The depths of time are so misty that we don't even know if viruses ruled the RNA world or were evolutionary detours from primitive cells relying on RNA.

Whatever happened back so long ago, viruses were central to life's evolution. They even became parts of other organisms. Plasmids (mini-sets of bacterial DNA) are one nifty example. They can replicate by themselves and carry important genes key to the host bacterium's survival, such as antibiotic resistance. Unlike the destructive relationship between their bacteriophage adversaries, bacteria have a mutualistic relationship with plasmids; a "you scratch my back and I'll scratch yours" agreement. In humans, viruses like HIV can actually fuse into our genome and stay there for life. Cool...but deadly. Actually, five percent of the human genome comes from viruses! So whenever you get a cold, consider that virus as a very, very, very obnoxious long-lost cousin!

However, the ability for viruses to sneak into our genome may actually be key to our existence today. Viral infections long ago that implanted their genes into your body might have given you a viral security system that helped force out harmful viruses before you are born. Some of these viral upgrades might've even beefed up our immune systems in the womb. Viruses can even treat cancer! Mysterious ancient texts describe smallpox, the flu, and other viruses momentarily stopping or even beating back cancer. This inspired generations of researchers, and after almost a century of trial and error the scientists finally bioengineered a virus that left healthy cells alone but gave no mercy for cancer cells. This virus, called T-VEC, is used to treat skin and lymph node tumors. Dozens of other candidates are being developed to treat cancer. These "good" viruses don't do any harm, though still cause the immune system to destroy them out of habit. This makes virus-based therapies a tricky business. But hang in there! Once this tech is improved, these "hero" viruses will be saving the lives of cancer victims like no other!

Keeping the Peace

Viruses help keep peace on this planet. Viruses attack their hosts and keep their population in check before they can push out other species. However, viruses must not go on a rampage and kill too much, or both they and their host will die out. This delicate balance defines why the world is so amazing and wonderful as it is today. Viruses were not meant to be evil. Mother Nature just made them keep organisms in line from day one, though it is tempting to imagine viruses as supervillains causing a global apocalypse.

Viruses somehow manage to be mind-boggling numerous yet avoid wiping out everything. If you lined up all the world's viruses, they would stretch ten million light years long (600,000,000,000,000,000,000 miles), more than the length of an average galaxy. Speaking of numbers, there are a lot of viruses in the ocean, too...ten million per drop of seawater to be exact. The oceans contain tons of bacteriophages, which prevent bacteria populations from going hog wild. That's a good thing! No one wants to see our entire ocean covered in dripping, green bacterial scum visible from space. Viruses keep everything the way it is, and remind us we still live in an RNA world.

It's a Wild World

Some other facts should be known about viruses before we leave. Viruses don't like to work too hard (like me on a Monday morning), so most can't infect more than one type of organism. Viruses that infect fungi don't infect plants, viruses that infect plants don't infect animals, and viruses that infect bacteria don't infect amoebas. Viruses stick to a branch on the tree of life and hang around there. Many are happy to just stay put in the animal kingdom.

And in the animal kingdom, there are always bugs. Think that you're out of the woods for getting viruses from insects? Well, you're out of luck. A huge group of viruses called arboviruses are transmitted by insect bites. One of them, the scary-sounding yellow fever, attacks the liver so much that it's unable to grind down bilirubin (the waste product of red blood cells), turning the skin yellow. The pigment is also toxic and causes massive tissue damage to your body if left untreated. Bouts of yellow fever don't seem bad at first; in fact, a patient may seem to recover after a few days in bed. However, severe relapses are common and, without treatment, frequently lead to death. Only decades of painstakingly draining mosquito-infested swamps and puddles, installing netting, and doling out vaccines have trounced yellow fever. But this virus is still popping up in areas with poor sanitation. Yellow fever's cousins—the flaviviruses, which include the notorious dengue fever virus—are also hard to shake off. While the bugs don't get sick if they contract these viruses, other animals do. These pathogens have been around for millions of years before humans showed up, and then they promptly spread to us.

A word about fish and viruses. Fish are the most ancient vertebrates ever and first appeared around 525 million years ago. The first fish didn't even look like fish; they looked like legless tadpole-worms! So it should come as no surprise that viruses that infect fish don't really infect humans, even though all vertebrate species have evolved from fish. Our slimy amphibian ancestors waddled onto land just too long ago. This saved us from contracting (most) diseases roaming the high seas. But on land, viruses can easily jump from species to species. The chikungunya virus, western and eastern equine (scientific name for "horse") encephalitis virus, and the West Nile virus can infect humans, monkeys, apes, other mammals, birds, reptiles, and even amphibians. They're also all spread by bugs. Argh! Mosquitoes and ticks are everywhere!

They're Here to Stay

Despite the deadly diversity of the viruses above, they all have one thing in common: they are all RNA viruses! Well, this isn't surprising. A LOT of mutations happen in RNA. Whenever a virus replicates, one out of every 100,000 "thorns" on the RNA will mutate. This may sound small, but considering viruses replicate extremely quickly, small mutations add up. This allows for incredible diversity: viruses can infect anything from alligators to zorillas (type of skunk-weasel). For example, the Hepatitis C virus attacks the liver. However, its RNA is so garbled and glitchy (that's actually a good thing, for it can become more transmissible and more virulent—or harmful—quicker) that its genetic material soon becomes ten percent more different from that of its original strain. That's more than the genetic difference between a human and a chimpanzee, and even the genetic difference between a human and a cat! This rapid evolution is the reason why so many viruses are so hard to shrug off.

But should viruses really take all the blame? Take Ebola, for example. It's so deadly to humans that it infects the bloodstream and shuts down the entire body. However, it originally infected bats in the jungle. These bats carry Ebola with them, yet the fuzzy fliers do just fine. Ebola, though it has such a bad rap with humans and primates, would never hurt a hair on bats. They evolved peacefully together; Ebola just spilled over from bats to us and our fellow primates. Mother Nature never knew it would cause a panic; it was just trying to improve one of its trillions of works of art. Evolution starts out with a blank canvas, and sprays paint everywhere to create a picture. Ebola is just one of the many blemishes on this otherwise beautiful picture (hey, for a piece of genetic graffiti, our natural world is quite something).

Bugging with Us, Living with Us

And as our journey comes to an end, we come to an important conclusion. RNA viruses are tricky little ninjas, thanks to their adaptable RNA. They quietly replicate in a host and cause no harm, violently spread, or somewhere in between. But beyond copying themselves like evil printing presses, viruses play a crucial role in keeping our ecosystems stable and enhancing genetic diversity. Though it seems like we've taken over the planet, Mother Nature warns us not to be too cocky with her vast army of pathogens. Humans will always need to deal with viruses. It doesn't matter if we settle on Mars, fight an intergalactic war against gigantic cat-like aliens, or become stuck in a spaceship wandering the far reaches of space for eternity. No matter where we go, viruses will always be with us. They are part of us. They were here first and will be the last ones standing. They have been bugging other organisms for thousands of millions of years. But don't feel too bad the next time you are sick in bed. You're just experiencing a part of evolution!

CHAPTER 4

EGGSTREMELY TROUBLESOME

Y OU MIGHT'VE WONDERED WHY YOU GOBBLED UP LESS SCRAMBLED eggs with your slab of bacon for breakfast in the early days of 2023. You might've also heard your parents stood in long, snaking lines to buy a carton of eggs. This egg shortage was the result of a deadly flu outbreak that made thousands of chickens bite the dust. The poultry-killing flu responsible for the carnage is dubbed HPAI, or Highly Pathogenic Avian Influenza. Pathogenic, if you haven't had the pleasure of knowing, means a disease causes severe symptoms in its host.

There are many strains of HPAI. The strain that brought the demise of the chickens is called H5N1, a minimalistic yet threatening name like the code number of a stormtrooper. H5N1 and its influenza cousins spread through birds' saliva, mucus, and feces. This makes H5N1 especially difficult to contain and corral, like rampaging bulls on a cattle drive. Farmers besieged by H5N1 are often forced to kill all their poultry in a process known as mass culling to prevent the virus from spreading. Plus, H5N1 doesn't just wallop chickens; the virus infects deer, bears, sea lions, and even dolphins. H5N1 only infects humans a handful of times each year, but infections can be severe. Plus, H5N1 changes very, *very* fast. The flu that cuts down your omelet supply might be infecting *you* in the future!

The Scoop on Antigens

But aside from grouchy chickens, what's up with the flu? Why do we need to get flu shots every year? The answer lies within the virus itself. The flu mutates rapidly each time it replicates. No, the virus doesn't shoot lasers from its eyes or grow taller than a skyscraper. Quite the opposite. The changes in flu viruses seem trifling, but are important. You've probably seen those spikes on most depictions of viruses that make them look like cacti. These spikes are actually viral proteins called surface antigens. Surface antigens are a viral barcode of sorts used by your immune system to recognize pathogens, allowing your white blood cells to gun down the invader with microbe-killing antibodies. However, the flu can sneakily shapeshift its surface antigens, causing previously-made antibodies to harmlessly bounce off the virus. This is why you can get the flu many times each year. The body fights the flu by creating new antibodies, but the virus still gives you a tough time. This is called antigen drift. Because of this, I caught the flu while writing this chapter!

A much more sudden genetic mutation occurs when the flu virus shuffles its genetic info and develops completely new surface antigens. This is called an antigen shift. For a good example about the devastating consequences of an antigen shift, let's time travel back to 2009. A swine flu known as H1N1 tore across the globe, killing countless pigs and thousands of humans. The pandemic sprung up so abruptly that healthcare workers worldwide were overwhelmed with millions of patients. Thankfully, vaccines were issued to stave off the virus. Whether antigen shifts and drifts spark nasty pandemics or merely make your nose water, both cause new variants of flu to gallop across the world each year, hence the annual flu shots.

Mix n' Match

The offspring of a Labrador retriever and a poodle is a fluffy labradoodle. Breeders often mix two breeds of dog, creating interesting results. Like dogs, the flu hybridizes too! When two different strains of flu attack a cell, they attach themselves to the cell and inject their RNA. The cell is "hacked," and forced to manufacture virus parts. This is when the (deadly) fun begins! The flu has eight different pieces of RNA genomes. Each piece is called a genome segment. When a cell manufactures two different strains of viruses, RNA segments are sometimes swapped, creating a viral Frankenstein! This process drives the deadly cycle of antigen drift, and even small antigen drifts can lead to potentially catastrophic results.

In tests on mice (popular test animals due to their startling genetic and internal similarity to humans), when the H3N2 human seasonal flu virus swapped a *single* segment with the H5N1 flu virus, it created a highly pathogenic strain, which was much, much deadlier than any flu we've seen before. In simulations, this lab freak of a virus infected both birds and mammals, producing a virus uncomfortably similar to the notorious H1N1 flu variant that claimed tens of millions of lives back in 1918. If this simple yet disastrous reassortment could happen in a lab, it could happen in nature. (Gulp!) But with a couple of innovative vaccine ideas, we could shield ourselves from the brunt of another pandemic. I guess it's Godzilla vs. King Kong...except on a much smaller scale between humans and the flu!

Speaking of strains of flu brings up the topic of flu types. There are three main types: Type A, B, and C. Types A and B bring on days of miserable sniffling in bed and cause death. Type C is gentler than its savage cousins and is left in the shadows. In this chapter, I will focus on Types A and B. Type A infects a menagerie of animals, including migratory birds, farmed poultry, cats wandering the streets, seals at the aquarium, and even whales wandering the

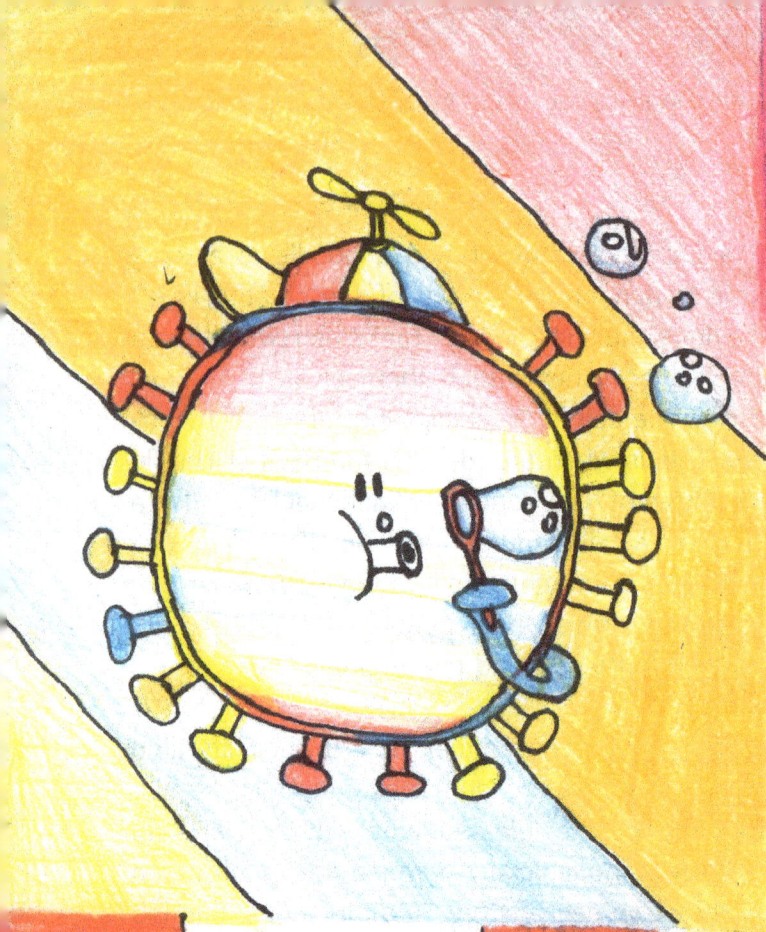

blue briny. Type B is mainly seen in humans, but occasionally pops up in seals.

There's an important difference between a *type* of flu and a *strain* of flu. A strain of flu is more specific than a type of flu. Each type, as previously mentioned, gives rise to all sorts of different flu strains through antigen drifts and shifts. To help you remember the distinction, each strain of flu has the stormtrooper-esque number setting it apart from other flu strains in its type. Sort of like Pokémon. For instance, the most famous strains of Type A flu are H1N1, which kicked off the 1918 flu pandemic, and H3N2, which kicks off nasty and potentially deadly flu seasons. Type B flu strains, on the other hand, don't cause pandemics because they're mainly exclusive to humans and don't get a chance to shuffle their genetic info.

Disney's Dilemma

I would like to give a special mention to Walt Disney for surviving the infamous 1918 flu pandemic. Back before Disney was drawing the lovable Mickey Mouse and cantankerous Donald Duck, he wanted to be a soldier in World War I. During his military training, Disney contracted the flu. He soon bounced back and served in the Red Cross, where he saw victims of the flu strewn across the battlefield. When he finished his service, he was tougher, more dignified, and ready to return home.

The Purple Death

The 1918 H1N1 flu variant is also called the "Spanish flu." But why Spanish flu? The H1N1 virus didn't originate from Spain; the moniker Spanish flu came to be because Spain gave detailed reports on the 1918 pandemic. The Spanish flu probably originated in the US. Spanish officials and doctors protested the name of the flu as not only a red herring, but also it stigmatizes their people. Unfortunately, the name stuck.

The 1918 pandemic was an agonizing experience. The H1N1 flu was even nicknamed the "purple death" because it killed other-wise healthy, beefy young people. Toddlers and the elderly weren't spared either. The flu infected 500 million people, a third of the world's population back then! It wasn't an easy time. Communities quarantined themselves. Schools and gyms closed down. People wore handkerchiefs (hankies) and cloth masks over their mouths. However, the virus easily slipped through these crude masks, ren-dering them useless. (Modern masks, on the other hand, have a much tighter mesh that is more effective for blocking viruses.) Handwashing was an important virtue during the pandemic. People shunned touching public items, such as library books.

Despite the world's best efforts to thwart the flu, society in the 1900s lacked treatment for the virus. Today, there's an arsenal of cough drops, flu medications, and vaccines to flatten the flu. But in the olden days, people were still scratching their heads about the flu. Microbiology was an obscure topic. There were no Walgreens or Rite-Aids offering flu shots. It's hard to believe vaccines are now taken for granted only 100 years later. Also, people weren't very big on hygiene. The luxuries of a flushing toilet and running water were exclusive to the wealthy. There were no sophisticated tests to sniff out the flu, nor were there any good medicines. Doctors diagnosed the flu with, at most, educated guesswork. A combina-tion of unfortunate circumstances made the 1918 flu super-duper deadly.

People tried everything to ward off the virus. They boiled chili peppers to "scare away" the flu, believing the virus was an evil spirit. They scattered onions all around the house to "soak up" the flu. Some even took laxatives to expel the flu out of their bodies by, erm, going to the bathroom. (This was a very bad idea. Laxatives made flu patients lose precious water and, back in 1918, contained toxic heavy metals like mercury.) Hospitals overflowed with suf-fering patients. Temporary hospitals were erected, but they didn't even make a dent in the number of flu patients desperately needing care. The H1N1 flu attacked with such ferocity that people died in a matter of *hours*, something unheard of today. The 1918 flu didn't just pack a punch of high fevers and coughs. Some people turned blue at their fingertips and even bled from their eyes. Nurses did everything they could; though they were swamped by casualties, these gallant medical workers kept caring for their patients. Graveyards were crammed with dead bodies. Some families even dug their own graves.

Like some terrible tsunami, the pandemic was not just one single wave, but a series of waves. The first wave served as an alarm bell after it sprung from the tail end of World War I. The second wave that began in the middle of fall was the worst. The third wave hit in the dead of winter, with howling winds bringing along the flu. However, there was hope. The appalling conditions during WWI that enabled the flu to spread so rapidly finally ended. The crowded barracks and ships were the culprits. Eminent scientist Sir Arthur Everett Shipley declared good ventilation and exposure to sunlight helped patients bash the flu. Open-air hospitals that gave patients plenty of fresh air and big bowls of food were a success story. The combination of fresh air, sunlight, and delicious food helped more patients survive in the open-air hospitals than in the crowded indoor hospitals.

After the third wave of the pandemic doused the world in misery, the clouds of despair parted, and infected victims slowly gave way to healthy bodies. The pandemic burnt itself out in the summer of 1919, either killing or granting immunity to so many people across the world that it had no place else to go. While many different varieties of H1N1 are still a threat, the 1918 variant is extinct. Humanity learned a stark lesson...or did we?

Revenge!

Another flu pandemic whaled humanity in 1957. It emerged when a strain of bird flu mutated and infected humans in Southeast Asia.

It killed more than a million people worldwide. This pandemic was bizarre: some people caught a cough and stuffy nose, while others developed lethal symptoms such as pneumonia. This startling variation occurred because some people already developed antibodies against similar strains of flu. This pandemic was dubbed the H2N2 pandemic. Fortunately, we were ready. The rapid development of a vaccine and various antiviral drugs curbed the deadly pandemic. Today, H2N2 only circulates in non-human animals. H2N2 is probably sorry it messed with us!

There was another flu pandemic in 1968, but us humans became champs at grappling with the flu. The 1968 pandemic was born from the death throes of the 1957 pandemic when the H2N2 virus mixed with another bird flu virus, creating a new strain of flu called H3N2. (Revenge!) H3N2 was first found lurking in the US in September 1968. There were "only" one million fatalities, much better than the fifty million deaths of the 1918 flu pandemic. Despite the carnage, medical advancements protected humanity. Most people who died were over sixty-five or had weak immunity. The H3N2 virus still exists today, albeit in a less deadly guise, and causes a handful of nuisances every year.

The H1N1 virus returned to the world's doorstep in 2009. This pandemic began when many unfortunate pigs in Mexico were infected with a novel yet sinister swine flu strain. This strain, named the H1N1 2009 pandemic strain, is still the most common flu strain today. (Remember the swine flu earlier in the chapter?) The first recorded human victim, a boy named Edgar, fell ill in a series of outbreaks in La Gloria. His family (correctly) blamed the crowded, pestilent pig farm near La Gloria for his illness. He recovered after a few days, but created a complicated legacy. A statue was erected in his honor, but Edgar was discriminated against at school due to the myth he was patient zero, or the person who started the swine flu pandemic. Though the first *recorded* case of swine flu occurred in the La Gloria pig farm, we still have no idea where the virus originated. Activists also started trying to shut down the pig farms, claiming that the farms contaminated the environment with slurries of noxious chemicals.

If the hullabaloo from La Gloria wasn't bad enough, a worldwide pandemic was underway. But despite the quick actions of medical authorities, this H1N1 flu was so different from other flu strains that a new vaccine couldn't be made in time. The vaccine was created in November 2009, after the second wave of the pandemic had taken its toll. But curiously, the elderly had an advantage in this pandemic. Most of them had been infected by other H1N1 viruses and possessed partial immunity. Children and young adults, on the other hand, didn't have their immune systems calibrated against the swine flu.

Fortunately, the swine flu only resulted in about half a million deaths worldwide. A lot less than the two- to seven-million lives medical experts expected the virus to claim! The confusion shrouding the beginning of the pandemic caused hysteria. Emergency disaster plans were set up, and people rushed to make vaccines. But compared to previous pandemic-spurring viral villains, the H1N1 swine flu was much less pathogenic—a lucky break for humanity!

Not Just a Seasonal Sniffle

Why is the flu—usually thought of as a seasonal sniffle—so deadly? The virus itself only causes cold-like symptoms; the real killer is the immune system's sometimes life-threatening response to the virus. Normally, the flu invades lung cells and the immune system destroys these infected cells. This process typically goes okay. However, the immune system occasionally goes into overdrive and destroys too many lung cells. This results in hypoxia, a fancy, tongue-tying word for suffocation, which can lead to death. Others die because of bacterial/fungal sidekicks helping the flu do its dirty work, such as the rascal *Streptococcus* (featured in Chapter

9). These infections are known as secondary infections and are frequent dangers during the flu season.

Outfoxing the Brainiacs

Let's move on to a less depressing but equally important topic: the flu vaccine. You might think the vaccine just gives you a sore arm. However, the prick of a needle has saved millions at risk of developing serious flu symptoms. In other words, the only reason why you think the vaccine is annoying is because it keeps you alive to think so! However, the business of making flu vaccines is tricky. Unlike other vaccines, which focus their efforts on just one or two viruses, the flu vaccine has to be constantly tweaked to keep up with new strains of flu. Doctors and scientists must literally predict which strain of the flu will sweep across the globe each year. Sometimes, these folks are right and a cushy flu season follows.

But once in a while, even big-brain scientists are outfoxed by viruses. Sometimes, a totally different strain of flu jumps from continent to continent like a bolt out of the blue. Even though the virus might be a strain that the vaccine isn't completely equipped to fight off, the vaccine can still protect you against contracting a severe infection. A flu vaccine is like an umbrella shielding you from a rainstorm. You still get wet by stomping in puddles or when the wind blows rain in your face, but most of the rain is deflected by the umbrella. While some flu strains create bigger "rainstorms" than others, your vaccine "umbrella" still helps you avoid being soaked. The vaccine also reduces your chances of coming down with the flu by forty to sixty percent. A sore arm is worth it if it means not spending your winter break in bed with a raging fever!

How did the flu vaccine come to be in the first place? It's a long story. At first, the flu was believed to be a bacterium. It didn't help that German scientist Richard Pfeiffer isolated bacteria in the noses of flu patients, which he called *Bacillus influenzae* in 1892. When the 1918 pandemic hit, people took antibiotics instead of antivirals, which were unheard of back then. Antibiotics work by piercing the cell walls of bacteria. Viruses don't have a cell wall, which is why you don't take antibiotics when you have a cold. Though microscopes had been around for quite a while, electron microscopes powerful enough to allow us to see viruses weren't invented until 1931.

The flu vaccine wasn't developed until after the 1918 pandemic. In 1933, scientists C.H. Andrews and Patrick Playfair Laidlaw made a major breakthrough: they discovered the influenza virus from throat swab samples. A few years later, Jonas Salk (who also invented the polio vaccine) and Thomas Francis made the first experimental flu vaccine. It was, and sometimes still is, made from deactivated viruses. These "dead" viruses don't cause any harm, but serve as target practice for the immune system. In the same year, scientists also discovered Type B flu. In 1942, a new vaccine was made to protect people from both influenza A and influenza B. After testing, the vaccine was given to soldiers in the military in 1945.

But scientists weren't done yet! In 1947, scientists realized existing flu vaccines didn't quell flu transmission. New strains of flu kept popping up every year. Then a group of scientists had a brainwave: why not track down strains of flu to make better vaccines? These scientists created organizations monitoring globetrotting flu strains, including the Worldwide Influenza Center and the Global Influenza Surveillance and Response System (GISRS), which were established in 1948 and 1952, respectively.

Cooking up a Vaccine

With all this history aside, I would like to point out when you receive the flu shot, you may be injecting chicken eggs into your body. Sometimes. While doctors don't stick literal bird eggs into you, the flu vaccine is traditionally made by "farming" flu viruses in eggs. Pork gelatin (a jelly derived from pig tissue) is also used to grow flu viruses. The viruses are then killed to be used as a vaccine. These vaccines do a good job of keeping you safe, but there's a catch. Human flu strains can infect birds but must mutate to infect and grow in chicken eggs. This makes these types of vaccines slightly less effective.

People with egg allergies usually get egg-based flu vaccines without any problems because there's very little egg protein in the vaccine. But for the sensitive folks out there, stranger options are available, such as vaccines made from viruses cultivated in dog kidney cells. Perhaps the weirdest flu vaccines are made from viruses harvested from fall armyworm cells. (For those of you who like bug collecting, fall armyworms are actually pesky caterpillars that munch on crops.) A trendy new vaccine relies on monoclonal antibodies—or lab-made antibodies—to clash with the flu. Egg-based vaccines, despite having kicked off the vaccine industry, will be a thing of the past.

Seek and Destroy

Special drugs called antivirals also hunt down the flu. Whenever your cells are attacked by flu viruses, they're corrupted to manufacture virus parts. To get inside a cell, flu viruses hijack your cells' receptors, which are normally used to receive chemical signals. Some antivirals deny access for flu viruses to enter the cell by plugging up the receptors. Some prevent the virus from copying its genome, functioning like a paper jam in a printer. Some even join with the virus's RNA to stop it from replicating. Other types of antivirals prevent viruses from crawling their way out of a cell. Flu viruses use a wacky-sounding protein called neuraminidase to blast their way through a cell. Certain antivirals bind to neuraminidase (hope this word isn't on your next spelling test) and prevent the virus from being released. This is why you might feel down if Mom gives you antivirals—this type of antiviral doesn't kill, it traps. It's up to your body to act and stop the tide.

The Waltz Goes On

The flu has a hidden dark side. It caused massive casualties in four major worldwide pandemics in the past century and continues to take the lives of 28,000 people annually. Fortunately, we've developed vaccines and antiviral drugs to stave off the virus. But the flu is still adapting, still changing. This is a classic global evolutionary race, and it has no intention of stopping. It's a never-ending war between humans and viruses. However, along the way, we've made miracle cures and other landmark medications. This isn't just a dramatic, desperate melee for dominance over our planet... no, it's a deadly but graceful dance between humans and nature.

CHAPTER 5

MOLD, GLORIOUS MOLD

IF YOU HAPPEN TO EAT MOLDY BREAD WITHOUT NOTICING (I'VE accidentally eaten raw pizza before, okay?), you might realize that you didn't get very sick. You might've gagged or acquired a stomachache, sure, but you didn't end up in the hospital. One of the critters to blame for spoiled food is *Aspergillus*, one of the most common molds on the planet. Look at the particles in the soil, the air, the petals of a flower, between the cracks of your sofa, and on your remote control under the microscope and you will find this little fungus. All of the many members of the *Aspergillus* family tree are quite harmless, that is, if you are healthy. Your immune system deploys an army of handy-dandy white blood cells to devour mold spores, leaving no chance for mold spores to germinate and cause infection.

However, that is if you have healthy lungs. Twenty out of the 185 strains of *Aspergillus* can cause extremely nasty symptoms in people burdened with asthma or allergies. The others can trigger coughing, itchy eyes, a stuffy nose, and dry skin. Fortunately, most of the time these reactions aren't very harmful. Every day, mold spores, pollen, and other nasties try to kill my nose, especially in the spring! However, sometimes these allergic reactions can cause *hypersensitivity pneumonitis*, a showy way of saying that someone's lungs are severely inflamed. This condition, unfortunately, can also be triggered by particles other than mold.

But before we continue on this musty, dusty journey, I want to make the distinction between mold and other fungi. Molds like *Aspergillus* resemble dandelions with branching stalks and reproduce via spores. They languish in nice, humid environments. Another type of fungi are yeasts. Yeasts reproduce by budding and look like beach balls. Mushrooms are similar to molds, but are visible to the naked eye and have complex organs like gills. Gills store spores and protect the delicate parts of mushrooms. All of these critters can be dangerous. Molds and yeasts can wreak deadly infections, and mushrooms can secrete toxins that can destroy the liver and kidneys. But mostly, mold is a multicellular (made of many cells) fungus, while yeast is usually unicellular (made up of only one cell).

Fungal Friend

But not all molds are bad. *Aspergillus* plays an important role in recycling nitrogen and carbon back into the environment. Some even produce cancer-fighting drugs and treat wastewater. It isn't the only "nice" mold out there. The spicy, tangy flavor of blue cheese is generated by a friendly mold called *Penicillium roqueforti*. If the name *Penicillium* sounds familiar, you wouldn't be surprised to hear the lifesaving antibiotic Penicillin is derived from *Penicillium!* Some molds are even used with yeasts to brew wine and beer. Like it or hate it, mold is a secretly beloved food around the world.

You'd think these obscure species of mold with tongue-tying scientific names are all boring, alien fuzz or extraterrestrial slime. However, *Aspergillus* mold creates a riot of colors beautiful enough to impress an artist. Species of *Aspergillus* such as *Aspergillus clavatus* cover the cool colors, such as blue-green and yellow-green. Some can also adopt more drabish colors, such as *A. fumigatus*, which can take on a bluish tint, and *A. nidulans,* which possesses an olive-colored hue. *A. terreus* is a lovely cinnamon shade. Not yet impressed? *A. versicolor* is the most flamboyant of all. It begins as a lovely shade of milky white, then erupts into a dazzling array of colors including brown, yellow, lime, reddish violet, and pink! Never thought a mold could be so pretty, huh?

Fungal Foe

However, when there's the good, there's the bad and the ugly. Many strains of *Aspergillus* can wreak havoc on the body. They are also members of the Superbug Club. Superbugs are microbes resistant to antimicrobial drugs, not gigantic insects, and they are a massive problem to the medical industry. Far from being just an annoying mold that destroys bread and flourishes on strawberries, *Aspergillus* has an arsenal of attacks and allergic reactions. As a common allergen, *Aspergillus* frequently causes allergic sinusitis. This allergy evokes a headache, stuffy nose, and a serious itch, but it is otherwise harmless.

A more powerful trick up *Aspergillus*'s sleeve is Allergic Bronchopulmonary Aspergillosis, or ABPA. With such a zany and eccentric name, ABPA sounds dangerous, but it is merely an allergic reaction that assails the lungs. ABPA cases kick up bouts of coughing and wheezing, labored breathing, and a fever, but they are usually quite trifling. These allergic reactions are the result of the immune system overreacting to an otherwise innocent "threat" and cause minimal damage to the body.

Aspergillomas signal the battle against *Aspergillus* is heating up. Aspergillomas sound a bit kooky but should be taken seriously. They are respiratory infections that occur when *Aspergillus* colonizes the nose and lungs. Aspergillomas are informally known as "fungus balls" due to their habit of forming disgusting globs of dead mold and mucus (gross). A fungus ball whips up a horrible cough, weight loss, fatigue, and can make it difficult to breathe.

Some devastating infections can occur when *Aspergillus* really worms its way into the body. Chronic Pulmonary Aspergillosis damages the lungs badly and Invasive Aspergillosis enters the blood, causing an infection in organ tissue. And the worst of them all is azole-resistant *Aspergillus*. Azole might sound like a rare vegetable found in a video game; however, it is actually a common antifungal medication used by doctors across the globe. All of these *Aspergillus* infections can spread to critical areas of the body, such as the heart or the brain.

Three Cheers for Azoles

While *Aspergillus* seems scary, it can be busted. Chest X-rays can easily uncover even the most elusive fungus balls. Small fungus balls go away by themselves, so treatment isn't generally needed. If a fungus ball stays put or continues to grow, a cast of antifungals are brought out to save the day.

This brings us to the drugs called azoles. We gave them a quick overview earlier, but I think it's time to give them the spotlight. Despite their quirky name, azoles have a surprisingly simple chemical structure. Azoles consist of a pentagon made of three carbons, a nitrogen, maybe some hydrogen, and in some cases, sulfur. It makes a wonderful antifungal. It works by stopping the production of ergosterol, one of the building blocks of a fungus's cell membrane. A fungus's cell membrane is like its skin. As you can imagine, things get pretty gruesome when your skin gets ripped off. These drugs are an excellent example of a simple thing making a great contribution, for azoles can kill a wide range of fungi, from yeasts to molds. It's a wonder drug! However, long-term use of azole poisons the liver and messes up your hormones, scrambling your mood and sleep. It's no coincidence azoles are toxic: fungi are much more closely related to animals than bacteria. This often makes antifungals harmful to both its recipient and its victim.

Surgery sometimes works better than antifungals. Large fungus balls can often cause bleeding when lodged in the lungs but removing them often solves the problem. If the bleeding continues, there's a wacky process called embolization to staunch the blood. Embolization is where a doctor inserts a tube guided through an artery to deliver a liquid that slowly hardens and forms an artificial scab. This is a great way to stop bleeding, but the bleeding may start again.

Warm, Wet, and Wild

But why is *Aspergillus* such a microbial menace? The immune system normally sends white blood cells to quickly eliminate mold spores. Unleashing its full power on a healthy person, *Aspergillus* can only create a harmless blob of infected tissue. However, people who are touchy with allergies or have asthma are sometimes unable to ward off the infection. This allows *Aspergillus* to flourish in the lungs. Lung tissue is warm and wet, a perfect environment for mold! If continued unchecked, *Aspergillus* spreads through the lungs like wildfire and reaches the alveoli. Alveoli are air sacs in the lungs that help transfer oxygen to the blood. These funny little guys look like grapes, but they play a vital function in the body.

Alveoli are terrific places for *Aspergillus* to grow. The body will try to defend itself by producing mucus full of white blood cells that try to fight off the mold with antibodies. If a colony of mold continues to grow, it will eventually cause bleeding, which the body tries to resolve by sending out blood-clotting protein called fibrin. Eventually, a fungus ball, or aspergilloma, forms. This disgusting result of the battle between fungus and man is a sphere of fungus fibers, scabs, blood, and mucus. If a patient is lucky, aspergillomas disappear over time. However, sometimes an aspergilloma won't go away and will continue to grow, causing widespread bleeding. If a fungus ball grows too big, it can suffocate you. Yikes!

However, *Aspergillus* can also use the lungs to start a bad case of invasive aspergillosis. (Remember the bad guy we went over earlier?) It can use the thin blood vessels in alveoli sacs to jump into the bloodstream and travel to other organs. This can cause hemorrhage—or fatal bleeding—in the lungs, inflammation of the airways, and organ failure (shock). The digestive system can also be pockmarked with holes in the stomach (ulcers), appendix infection, and attacks of pain. In the most extreme cases, *Aspergillus* can crawl into the brain and lead to neurological damage. This begins with delirium, which is a shift in attention and cognition. Later, blood vessels in the brain can burst and cause a stroke, killing the unfortunate victim.

Besides causing damage to people, *Aspergillus* can also have some nasty effects on animals. It can kill cattle and sheep fetuses before they are born. Since *Aspergillus* grows on plants, spreading across grain used to feed the animals is an easy task. On plants, it can secrete aflatoxins, especially *A. flavus*. These toxins can make poultry that ingest this grain, such as chickens, severely ill.

Killing the Cold-blooded Killers

The raw power of *Aspergillus* and other fungi may have also been partially responsible for the extinction that wiped out the dinosaurs. When the giant asteroid struck, global temperatures plummeted. Plants, which rely on sunlight to make food, almost disappeared from the face of the planet. This caused a massive fungal bloom. After all, fungi love cold temperatures. Reptiles are less resistant to fungus because they are cold-blooded. The sheer cold of the extinction made reptiles much more vulnerable and made it harder to produce fevers, which usually kill fungal infections. Though some dinosaurs had some control over their body temperatures, their weight-reducing air sacs—also found as air pockets in the skulls of smaller dinosaurs—were great targets for fungi. Even if dinosaurs survived the initial fungal outbreak, their eggs were also vulnerable. In contrast, mammal babies are safe and warm in/on their mother, while bird babies have a head start in being warm-blooded. This heat-related selection of "survival of the fittest" may be the reason why we are here in the first place!

Since fungi hate heat, they often target mammals with low body temperatures. Platypuses in Tasmania, an island off the coast of Australia, have been observed suffering from a mold called *Mucor* because of their low body temperature of ninety degrees Fahrenheit. (That temperature would be similar to a person suffering from severe hypothermia.) Hibernating bats, whose body heat pales in comparison to when they are out and about, are prime targets for the famous bat-killing fungus named *Pseudogymnoascus destructans*, or white-nose syndrome. This fungus causes the bats to wake up so they can kill the disease with a fever, then fall back asleep. This is very bad for the bats since they use up their fat reserves quickly when fighting an infection. If you are wondering of the reasons why we suffer from those irritating fevers when we are sick, fevers are essentially a built-in fungi-killer. It's one of the reasons why we haven't joined the fossil record with the dinosaurs!

However, global warming may change this picture. In the short term, it may be bad news for fungi as they are repeatedly slaughtered by warm weather. It also may mean more days lazing by the beach for people. But over time, fungi may develop heat resistance via natural selection. The tables could be turned, and our warm blood could no longer protect us from fungi as it did ever since we were mouse-like creatures scurrying around in the trees. There would be many more fungal illnesses ravaging not only humanity, but also hundreds of animal species. Fungi such as *Aspergillus* are pretty deadly when infecting immunocompromised patients, so if global warming causes mutant fungi to burst forth, we would all be toast.

Wheat Will Rock You

Speaking of disease brings us back to moldy bread. The reason why we don't eat moldy bread (aside from its disgusting appearance) is that *Aspergillus* is toxic. Remember the aflatoxins secreted by *A. flavus* that killed the chickens? These aflatoxins can also be deadly to humans when consumed since they can cause very bad cases of liver cancer. This is called mycotoxicosis. Exposure to small amounts of aflatoxin is usually not very deadly. So if you happen to eat moldy food, what's done is done. Just avoid it next time, and you'll be fine.

However, some people in developing countries can't access good food or don't have the luxury to store it for long periods of time. There have been many aflatoxin-related poisonings in history. One occurred in Kenya, an African country, in 2004. The outbreak was a nasty one. Out of the 317 people exposed to the lethal mold toxins, 125 died. It turned out that a bunch of corn was contaminated with a high dose of aflatoxins. Rainy weather followed by poor storage conditions made *Aspergillus* colonies erupt throughout the corn. Corn is a staple food in Kenya, and the mold grew on the damp corn. This happened again in 2005 and 2006, causing fifty-three more deaths.

Sometimes, aflatoxin outbreaks can hit the global level. Take the aflatoxin dog food outbreak that happened in 2005. More than seventy-five dogs died of aflatoxin infection, and hundreds more of man's best friend suffered from severe liver problems. The contaminated batches of pet food were shipped across twenty-two US states and twenty-nine different countries. When these batches were tested, sixteen were heavily contaminated with aflatoxins. Many pet food companies now test for aflatoxins in their critter chow.

Joining the Resistance

There's a much larger threat than aflatoxins, however: azole-resistant *Aspergillus*. Azole is also a choice of fungicide and insecticide. It's often used by farmers to keep greedy fungi and insects from snacking on their prized wheat. However, agricultural azole overuse has made nineteen percent of all *Aspergillus* strains resistant to azole. How? Simply by natural selection. There were a few strains of *Aspergillus* more resistant to azole than others thanks to mutations, and they survived being sprayed. As azoles were overused, each generation of *Aspergillus* began developing stronger mutations against azole. Eventually, some strains of *Aspergillus* became totally resistant against the drug. Today, surgery is often the only option when it comes to removing the bothersome fungus. Ouch!

Aspergillus is also easily misdiagnosed. Some of its symptoms resemble tuberculosis, and doctors easily fall for the red herring. *Aspergillus* is also similar to many other molds, even when examined meticulously under a microscope. Still, *Aspergillus* isn't completely invisible. CT and X-ray scanning are helpful for tracking down aspergillomas. Tissue tests—such as biopsies, which examine small samples of tissue—and blood tests for *Aspergillus* antibodies and fungal DNA are also a step ahead.

Fuzzy Friends Beware

Animals, from dogs to dinos, are susceptible to *Aspergillus*. However, *Aspergillus* attacks birds with a particular vengeance. Why? Birds have air sacs, big balloon-like structures that lighten their bodies so they can fly. These air sacs are ideal places for fungi to grow. They don't receive a lot of germ-killing mucus, nor do they have cilia (think tiny scrubbing brushes) like in human lungs. This makes it hard for the air sacs to get rid of the fungus, especially since they don't receive a lot of nutrients and oxygen from the blood. Birds have heterophils instead of neutrophils, the most common white blood cell in humans. Heterophils are less effective for preventing *Aspergillus* infection. Birds can suffer from shortness of breath and quickly lose weight. Often, they fluff up their feathers in a futile attempt at keeping themselves warm and comfortable. These brooding birdies are not the playful parrots you usually see in pet stores. And if you are feeling morbid today, the species of *Aspergillus* most fatal to birds is *A. fumigatus*. This species of *Aspergillus* causes up to ninety percent of aspergillosis-related deaths in birds. (*A. fumigatus* is also responsible for most of the mold infections in people.) The culprit of many *Aspergillus* infections in our feathered friends? A seed-only diet. Unlike birds that munch on fruits and veggies, picky eaters that only chow down on seeds are particularly vulnerable to aspergillosis. Seeds don't have much Vitamin A, which supports immune and respiratory health.

If Tweety skips out on salad greens and becomes infected, it's a rough, downhill ride. Birds are bombarded with scan after scan of X-rays to detect *Aspergillus* lesions. DNA tests scour blood and tissue for *Aspergillus* genetic info. Serology tests sniff out antibodies produced to attack *Aspergillus*. However, these tests often deliver negative results because aspergillosis often occurs in birds with weakened immune systems that struggle to make antibodies. If a vet decides to use a tracheal aspirate to check for fungi, a small amount of liquid is squirted into the breathing tube, or trachea, of the bird. The liquid, along with the infectious material, is then drawn up by a syringe for testing. But testing of tracheal aspirates often send out a false alarm: every living thing has a tiny bit of natural concentration of *Aspergillus* in their bodies. A better approach called laparoscopy shoves an endoscope—a camera specialized for viewing internal organs—down the bird's respiratory system to analyze infected material. (This is definitely still no fun for the bird though.)

Application of a variety of antifungals, inside and out, can be a lifesaver for our feathered friends. *Aspergillus* frequently grows and forms a disgusting slime on tissues. This slime is called fungal plaque. Surgical removal of these plaques is another possible step to recovery. However, the most important part of treatment is to keep the bird's immune system strong. This can be accomplished through hospitalization, gentle warmth, increased oxygen levels, inflammation-soothing medications, and ample feeding. Though

Aspergillus is a nasty, deadly pathogen for birds, let's hope all the birds that have aspergillosis out there can recover.

Fortunately, dogs and cats don't get this disease as often. Like humans, though, your fur babies can get unlucky and contract aspergillosis. Dogs often suffer nosebleeds and a runny nose because *Aspergillus* damages nasal tissue. Cats also develop the sniffles, but also contract some other symptoms not found in dogs, such as enlarged facial lymph nodes (which filter certain fluids through the body) and super-sized eyes. In rare cases, *Aspergillus* can spread throughout the body and cause widespread infection. In dogs, symptoms can vary widely. Vomiting, laziness, and floppy limbs are not uncommon. In cats, the situation isn't much better, with extreme loss of appetite, bouts of coughing, stomach trouble, and shortness of breath.

Certain scans, such as MRI scans that use powerful radio waves to create detailed images of the body, and CT scans that use X-rays and state-of-the-art computer technology, are very good testing options, but not always available. A more widespread but less accurate practice is nasal rhinoscopy. This process uses the same endoscope scenario to detect *Aspergillus* in birds. Even ultrasound is used to seek out infected tissue in cats when they suffer *Aspergillus* eye infections. Like a detective's sidekick, polymerase chain reaction testing—or PCR testing—snoops *Aspergillus* DNA in a sample and churns out copies of the fungi's genetic info, helping scientists better analyze the fungi's genetic material. (PCR is also used for COVID testing.) With the infection confirmed, samples are often taken with an endoscope and cultured in the lab to confirm if it really is this troublesome mold.

Cats and dogs both can be treated by inhaled, applied, and swallowed medications, though oral medications may need to be given for months. Inserting medicine-stoked tubes into the parts where *Aspergillus* is doing its dirty work is sometimes used. However, a more comfortable way to deliver similar results is to spray medication throughout a cat or dog's nostrils, sort of like how people sometimes use nasal spray to clear their nose. Even medicated bandages carefully placed into a pup's nose can help! Despite all of this treatment, dogs and cats often suffer chronic *Aspergillus* infections for up to five years. Disseminated aspergillosis, or when *Aspergillus* is spread throughout the body, can take an extremely long time to be cured. In addition, *Aspergillus* infections often pack a serious one-two punch: there are countless recurring infections!

There's Spore to the Story

Aspergillosis can be prevented by covering up when doing activities that kick up dust, such as gardening. There is a reason why we wear gloves when we dig up our sweet potatoes, after all. Caring properly for wounds by washing them with soap and cleaning them with antifungal cream can help prevent *Aspergillus* from slithering into cuts. Testing for early infection can help, especially when undergoing organ transplants. When you receive an organ, you generally have to take medicines that suppress the immune system to keep it from attacking the new organ. Dust masks and air conditioners are effective *Aspergillus* filters. Itchy noses are treated with doctor-prescribed medications.

If you want to make sure your pet canary stays healthy, clean its cage often to make sure dust and moisture are kept at a minimum. Feed it fresh, nutritious fruits and veggies, as *Aspergillus* often grows on musty food. (Make sure to Google a bird's dietary restrictions first.) Another good way to prevent aspergillosis is to make sure your feathered friend gets fresh air. Thankfully, most healthy birds are safe from *Aspergillus* unless they are exposed to a large amount of the fungus.

However, it is almost impossible to prevent *Aspergillus* exposure to cats and dogs. In fact, some breeds of pets are more vulnerable to *Aspergillus*. Because of their long noses, collies, greyhounds, and dachshunds (wiener dogs) are more susceptible to aspergillosis. Cats with thick, fluffy fur are also more vulnerable to aspergillosis. These cats are often unintentional dusters with their fuzzy fur, which traps *Aspergillus.* Since even healthy pets can develop aspergillosis if exposed to it in large amounts, you can still help by making sure that *Aspergillus*-rich material such as moss, hay, and poo is kept away from your pet. And make sure Fido doesn't roll around in feces, especially after you shampooed him so well!

Most of us have healthy immune systems, but that is no excuse to slack off on antifungal research when others are not as fortunate. *Aspergillus* antifungal research still needs more *oomph* into it. *Aspergillus* is also becoming drug-resistant. In fact, patients who are fighting azole-resistant *Aspergillus* are thirty-three percent more likely to die! Fortunately, this resistance is not widespread yet. However, we should not let our guards down. So many viruses, bacteria, and other fungi have become monsters because new medicines were not developed in time to treat them. We will need to develop more antifungal drugs to combat *Aspergillus*… before this fungus becomes totally invincible!

CHAPTER 6

SMELL THE GRAPES?

*P*SEUDOMONAS AERUGINOSA, A BACTERIUM WITH A MOUTHFUL of a name (say sooh-duh-MOW-nuhs-eh-roo-gi-NOW-suh), smells a bit like grapes. This bacterium also takes on the dull-green hue of grapes due to a pigment called pyocyanin, which causes the bacteria to acquire a cool glow-in-the-dark effect when exposed to UV light as a cool bonus. Like grapes, it is commonly found nestled in the soil. And like grapes, it needs lots of water to survive. (Just a little bit of trivia: wine grapes actually consume more water than other fruits.) *P. aeruginosa* is harmless enough, *usually*. Most people with healthy immune systems shrug it off without any issue. In fact, you might've caught *P. aeruginosa* before. Think back to the last time you had an ear infection. Chances are that it was caused by *P. aeruginosa*.

However, not all people have healthy immune systems to fight off *P. aeruginosa*, and this can be a real can of worms. Hospitals use all sorts of equipment to help patients recover, such as breathing machines, catheters (needles that inject medicine into the body), and humidifiers. Though bacteria frolic in sticky, soggy environments, humidifiers actually minimize the risk of airborne infections by using water to "push" other flying pathogens to the ground. But these damp pieces of equipment are favorite places for *Pseudomonas* to grow on. Since patients in a hospital are directly having these devices inserted into themselves, *Pseudomonas*

gets a free ticket into the body. In addition, most hospital patients aren't very healthy, meaning they can't shrug *Pseudomonas* off so easily.

A Tough Nut to Crack

Will antibiotics save the day? Probably not. *P. aeruginosa* is a superbug. A superbug is an extremely dangerous microbe that is resistant to many drugs. But *P. aeruginosa* is no ordinary superbug. Like an S-tier video game character, this microbe is impervious to almost every antibiotic on the market and can even swim around in hand sanitizer! One of its tricks of the trade is biofilm—a sticky, gooey bacterial sludge that's excruciatingly hard to get rid of. Ninety-four billion dollars are spent annually on treating infections involving biofilms in the US, and more than half a million people die from these gloopy messes every year.

But how does *Pseudomonas* build biofilm in the first place? The process is quite neat. Swarms of bacteria first use their flagella, or "tails," to attach themselves to a surface. *Pseudomonas* relies on iron to build biofilm. But other crucial components of biofilm can't be accessed in the environment alone, such as a protein called lectin, which is used to anchor the biofilm. To access these rare goodies, bacteria start chatting with quorum sensing. If you are wondering what the heck quorum sensing is, it's merely a mechanism that bacteria use to "talk" to each other. After a few "lucky" victims have been chosen, a few bacteria then kill themselves (ouch) so they can make lectin and other biofilm building blocks from their corpses. Sounds cruel? It is better for most of the microbes to survive with some small sacrifices than for all of them to die from an external force.

After the first blobs of biofilm are formed, mushroom-like towers completely made out of dead bacteria spring up from the ground. When conditions are right for the bacteria to spread, some bacteria bust a hole in these biofilm fortresses. These "chosen ones" inside the biofilm exit through this hole and fly all over the place, landing on new areas to colonize. Biofilm is a particularly useful "life hack" for bacteria, and it can be used both inside and outside a body. Biofilms also act like a force field for bacteria, sealing them safely away from the antibiotics, white blood cells, and other bacterial dangers.

P. aeruginosa can even share its biofilm house with other pathogens, such as *Streptococcus*, *Staphylococcus,* and *Candida*. However, *P. aeruginosa* bacteria are not the most lenient of hosts. They compete for limited resources inside the biofilm with *Candida* and *Staphylococcus*. And if *P. aeruginosa* is feeling grumpy, it attacks them with a plethora of chemicals that slow their growth. But *P. aeruginosa's* guests aren't afraid to fight back. They create chemicals that attract the body's white blood cells to chomp through the biofilm and eat *Pseudomonas*. *Streptococcus pneumoniae* in particular triggers *P. aeruginosa*. *Pseudomonas* and *Streptococcus* both live in the lungs, so when they compete for the same resources, things get nasty. These two are mortal enemies!

Porin' the Bad Stuff Out

Besides its annoying biofilms, *Pseudomonas* is pretty hard to kill on its own. Its cell membrane, which is actually made up of multiple membranes, is so thick that it is essentially a solid, portable steel wall. Any stray antibiotics that leak through this membrane are immediately transported out of the cell with specialized machines made out of proteins called efflux pumps. There are many kinds of these pumps, with weird names such as OprM, OprJ, OprN, OM, MexA, and MexX. Each of them specializes in "pumping out" certain kinds of antibiotics and other harmful chemicals, such as disinfectants or heavy metals. These miniscule machines are truly wondrous, despite being a huge nuisance to doctors.

Most bacteria have porins. They are funny little channels made of proteins found in the bacteria's membrane. Porins allow nutrients and other important molecules to pass through the cell walls of bacteria. However, the main caveat with porins is that antibiotics can wiggle through porins to attack the bacteria from the inside out. However, *Pseudomonas's* porins can close up to block antibiotics from entering the bacteria. So how does *Pseudomonas* get its food if its porins are all plugged while under siege by antibiotics? It turns out the porins that match the shape of antibiotic molecules put up the shutters, while the others have different shapes and are free to take in the bacteria's grub.

But the most peculiar trick up *Pseudomonas's* sleeve is that *Pseudomonas* can literally alter its genome in a matter of *hours*. If antibiotics have breached its cell membrane, efflux pumps, and porins, the bacteria can mutate extraordinarily quickly to develop even more defense mechanisms against the antibiotic. Because of its never-ending battle with antibiotics, *Pseudomonas* has effectively outwitted and outsmarted all of humanity's most potent micro-weapons! Perhaps one day in the far future, *Pseudomonas*— with its impressive adaptations and versatility—will be one of the last living things left on Earth.

Blood, Guts, and Gore

If you are intrepid enough to learn about the disgusting stuff that *P. aeruginosa* can do to the body, here comes the gore and action! We'll start from the head. Though not common, *P. aeruginosa* can cause very nasty eye infections, which paint the entire eye green with bacterial growth. *P. aeruginosa* is not a picky eater: it can digest keratin, which happens to be the stuff parts of your eye are made of. Not a lot of bacteria can do this! Keratin is also what happens to be the substance that makes up your fingernails, and *P. aeruginosa* can literally eat nails as well, all the while forming green scum as it goes. Burn wounds also happen to be targets of *P. aeruginosa*, being warm, wet, and having lots of exposed flesh. This can turn the entire wound into a mess of sickly yellow-green biofilm-infested flesh. Some parts of the skin near these infected wounds ooze out pus like lava from volcanoes. *P. aeruginosa* is one of the main reasons why burns sometimes heal so slowly. They are also one of the main culprits for the bloodstream infections that so frequently occur after these wounds.

Aside from hospital settings, *P. aeruginosa* can also be found lounging in dirty swimming pools and hot tubs. However, *P. aeruginosa's* preferred stomping grounds are very specific: in the lungs of people suffering from cystic fibrosis. Cystic fibrosis is a genetic disease that causes sweat, mucus, and digestive juices to become gooey and viscous. As you can imagine, this can become extremely uncomfortable and even life-threatening, especially when thick gobs of mucus clog up the lung. Thanks to handy modern technology, people who suffer from cystic fibrosis can live relatively long and healthy lives. But in the olden days, people with cystic fibrosis died by five years old. But why does *P. aeruginosa* like slithering around in mucus-infested lungs? It's because the mucus generated by cystic fibrosis-affected cells traps *P. aeruginosa* in the lungs, causing infection. The bacteria don't want to leave, either. After all, *P. aeruginosa* LOVES mucus!

P. aeruginosa can cause myriad horrible symptoms in people's lungs. Difficulty breathing isn't uncommon and is often accompanied by outbursts of green and/or bloody mucus. Fatigue occurs when the body is weak from fighting prolonged infections. When looking at the urinary tract, things are not much better. *Pseudomonas* can make peeing a painful business yet paradoxically make nature call 24/7. This bacterium makes urine turn green, cloudy, or even bloody. Worse, *Pseudomonas* evokes an agonizing case of hip pain when it infects the bladder.

A Shocking Matter

Once *P. aeruginosa* invades the body, it not only infects the internal organs but also the bloodstream. It produces a powerful toxin that can cause sepsis. This deadly condition caused by a blood-stream infection provokes widespread inflammation due to the body overreacting in an attempt to defend itself. This in turn can cause a cascade of nasty effects, such as blood clots and leaky blood vessels, and eventually organ failure (shock).

There are three stages of sepsis. The first is confusingly called sepsis. It is deceptively mild, causing a low fever and a racing heart. The second stage, called severe sepsis, starts an avalanche of unpleasant symptoms such as extremely rapid mood swings, attacks of unconsciousness, heart malfunction, chills, pain, and respiratory distress. The third phase is called septic shock. Widespread organ failure means patients suffering from septic shock are close to death. Only half of patients survive this stage. In fact, septic shock is so deadly that the chances of dying from this deadly condition rise by eight percent every hour without treatment!

Down the Drain

There is a way to stop *P. aeruginosa*; however, the bacteria are just too fast most of the time. This often happens because of one simple reason: people forget to wash their hands and don't bother to clean equipment. Surprised? Don't think you never forgot to spiff up your room or wash your hands before grabbing a snack! When a doctor or a nurse forgets to clean equipment with *P. aeruginosa* growing on it, the bacteria can be transmitted to another patient when the equipment is used again. Unfortunately, some of the equipment used by doctors is extremely hard to clean due to their sophisticated designs and structures.

Sometimes, it's the plumbing itself that needs to be cleaned. In hospitals, even the most seemingly benign places, such as sinks, drains, and toilets, can be great places for microbes to thrive and spread. After all, when you flush the toilet, you are basically launching germs onto your shirt. To you and me, they aren't a big deal. But to the patients already burdened by an infection, these microbes can cause massive damage. Wherever water is still or poorly disinfected, *P. aeruginosa* explodes into existence. Even the tiniest slip can allow *Pseudomonas's* biofilm to flourish and thrive.

Double (or Triple or Quadruple) Trouble

There have been many outbreaks of multidrug-resistant (MDR) *Pseudomonas aeruginosa*. One particular outbreak happened to a few people in Tijuana, a city near the US border in Mexico. They were practicing medical tourism, which is the process of seeking treatment in another area due to lower costs and convenience. They were infected while undergoing weight loss surgery. The surgical equipment wasn't sterilized properly, and a gob of *Pseudomonas* entered their bodies.

One of the victims of the outbreak had a stitched gash that would leak pus, and doctors had to cut the incision open to drain it out. Her condition grew so bad that doctors decided to use colistin—a last-resort antibiotic that causes kidney and nerve damage—to treat her infection. The antibiotic caused her face to go numb, so doctors had to stop the treatment. As a result, she had to clean the gash *herself* every day. Doctors did everything they could do, but the wound wouldn't heal. Fortunately, it didn't get worse either. She was luckier than most; another patient's immune system was overwhelmed so badly by MDR *P. aeruginosa* that he died. Outbreaks like this are not confined to Tijuana: other outbreaks of antibiotic-resistant *P. aeruginosa* occur across the world.

Another outbreak happened in Brazil. Twenty-seven people were infected with MDR *P. aeruginosa*. These sly bacteria had clambered onto surgical drapes (big cloths similar to those bibs you wear at the barbershop), surgical tables, and surgical drains (mechanical devices that remove infected fluid from the body). Most of the patients involved were exposed to equipment contaminated with *P. aeruginosa*. How did these wily bacteria get there?

Blame an open drain in the surgical room. The outbreak continued for a few years until the drain was sealed. Drains loaded with *P. aeruginosa* were blasted with air with each gust of wind, allowing the bacteria to gallop around the room. People did not discover that the drain was causing all the problems until two years later and after two people had been killed!

A Bacterial Brawl

Fortunately, there are various treatments being developed trying to break the almost impenetrable defenses of *P. aeruginosa*. When scientists put a metal called gallium into *P. aeruginosa's* biofilm, it collapsed, exposing the bacterium to the wrath of antibiotics. Medications containing this metal show some promise in treating *P. aeruginosa*. Chemicals that suck up iron from *P. aeruginosa's* environment are also surprisingly promising treatments. Iron is a crucial component in this bacteria's biofilm, and without it, *P. aeruginosa* can't play its trump card. Some scientists are even experimenting with tiny capsules that carry drugs through holes in the biofilm and attack the bacteria from the inside.

We could also eavesdrop on this bacteria's quorum. If we interfered with these bacterial "conversations" in the right way, we could prevent biofilms from forming in the first place. Researchers have also delved into potential biofilm dispersers that literally melt holes in *P. aeruginosa's* armor. However, all of these exotic treatments are quite toxic or disperse harmful bacteria. Drat! We're back where we started!

There is actually something that we can use right now to treat *P. aeruginosa* though. However, it is not a super-antibiotic or a high-tech chemical. It has been with us—and before us—all this time. It has been built for the very purpose of destroying *P. aeruginosa*, and as a bonus, is able to dissolve and penetrate through various types of biofilm. This is the bacteriophage. (We put this guy in the spotlight in Chapter 19.) A bacteriophage, or phage for short, has been the champion hunter of bacteria for billions of years. The classical phage looks like a spider drawn by a toddler, with a giant head fixed on top of a narrow shaft with spindly legs. Some phages, however, only have the head part. And a few very weird ones look like pieces of rope.

Phages kill bacteria like self-replicating ballistic missiles. The phage first locks onto its target with its legs after finding a receptor, which is used by bacteria to receive signals from other bacteria. Next, it injects its genetic material that is in its "head" into the bacteria with a sharp spike on its rear end. Then, it "hacks" the bacteria's genes to turn it into a bacteriophage factory. When enough phages are made inside the bacteria, they burst out, killing the bacterium and seeking more targets to destroy.

Bacteria are often portrayed as helpless victims to the phage. Not at all. Bacteria switch their receptors into different shapes to prevent phage entry. The phages then shapeshift their spikes to match the new receptor. Sometimes, bacteria grow flagella to move away from phage-infested areas. The phages then follow in hot pursuit. Even if phages enter bacterial cells, bacteria still have mechanisms to fight back. Bacteria can develop defenses that prevent phages from hijacking their genes, rendering the virus's attempts to multiply useless. Often, the phages switch the shape of their spikes again, using a disguise that the bacteria can't see through, to infiltrate their system again.

It's an endless war, but this method of killing bacteria is much more secure than antibiotics. If one phage fails, thousands of other strains of phages are available. If everything goes wrong because of bad luck, a new, promising phage can be discovered in a few weeks, or even a few days. In comparison, antibiotics take years to be developed—there are also a limited number of antibiotics to be found on the planet. Most of them are found in plants hidden in remote jungles or surreal creatures deep under the sea and often have to be tinkered around with so they are fit for human consumption.

Bacterial resistance to phages is still a problem, though less prominent than with antibiotics. However, people still worry that the wrong phage could attack the good bacteria in a person's micro-biome instead. That's why scientists keep a gigantic, online list of all the kinds of phages that are promising candidates for treating illnesses, so they don't end up having accidents. But to outstep the problem of phage resistance, doctors use a "phage cocktail." Phage cocktails are medicines prepared from a mixture of phages. If one phage fails, the others are there to back it up. Phage therapy is very promising, and if antibiotics become obsolete, phages will still be here to save our butts. Scientists have also found one very promising fact: if bacteria try to become more resistant to phages, they will be more vulnerable to antibiotics! Therefore, combining phages and antibiotics could also be a clever solution.

King of the Superbugs

The reign of *Pseudomonas* is a great example of the consequences of our own wanton mistakes. Sure, it seems like we have taken over the planet, but bacteria and other pathogens are controlling many of our actions. To microbes, we are marionettes in a puppet show. To break free of their puppet strings, we must be cautious and continue to develop new, effective treatments and prevent the king of the superbugs from shattering the peace!

CHAPTER 7

JUNGLE JAPES

Y OU ARE HIKING IN THE JUNGLE IN A LUSH RAINFOREST. YOU enjoy the twittering of the birds, the raucous hooting of the monkeys, and the faint trickling of raindrops splashing down onto the forest canopy. As you take a breather on a hilltop, you hear something else: a buzzing that sounds like the thrumming of tiny wings. You think it is a helicopter, but you realize that helicopters rarely come out here. You gaze at the horizon, then screech in terror: a giant swarm of mosquitoes is making its way toward their next meal (you). Panicking, you take out a can of bug spray and smother yourself in DEET. The black tide of mosquitoes diverts its path around you. You are safe from those pesky mosquitoes... for now.

Aside from saving yourself from being sucked dry (gross), you may have also avoided contracting malaria. This deadly parasite is transmitted by the Anopheles mosquito. As people once believed that the illness was caused by the rancid air of swamps and marshes, malaria is Latin for "bad air." This parasite is one heck of a disease. One estimate claims malaria killed half of the 108 billion people since the dawn of humanity! In World War II, the Nazis considered unleashing swarms of malaria-infected mosquitoes to bombard their enemies. Even today, half of the eight billion humans are at risk of contracting malaria, and the disease claims a child's life every thirty seconds.

Fortunately, many countries are free from the scourge of malaria and have been deemed malaria-free. According to the World Health Organization, or WHO, a malaria-free country has not had a malaria case for three years straight. From Albania to Uzbekistan, many countries garner the title of malaria-free. In developed countries, malaria is so rare that the disease is misdiagnosed as the flu and not treated seriously, which is a big mistake. But in sub-Saharan Africa where malaria is endemic (spread locally and continuously), people know when they have malaria, and they treat the parasite immediately once it rears its ugly head.

When you venture out from the tropics, malaria is not a common concern. Malaria cases pop up occasionally in airports in Florida and southern Spain each year, but this is because mosquitoes with malaria hitch a ride on planes to these parts of the world. Though malaria is usually restricted to the tropics, malaria cases have been spotted as far north as Norway. Unfortunately, the US is no longer malaria-free. In 2023, several *local* malaria cases sprang up in Texas and Florida. All of the patients recovered, but this may be a prelude of things to come. With climate change bringing increasingly sweltering temperatures, mosquitoes laden with malaria threaten to invade the US again!

Those Darn Mosquitoes

Malaria begins its dirty work as quickly as six to eight hours after a mosquito bite when the parasite invades the bloodstream. However, symptoms may not pop up for as long as thirty days after the parasite slithers into a human body. With a series of lightning-quick "attacks," malaria induces fever, vomiting, bouts of shivering, and squeezes out bucketfuls of sweat. Malaria victims usually show signs of recovery after each attack. But this can be deceiving. People infected by malaria are often assailed by multiple attacks, which leads to an avalanche of complications such as blood clots and organ failure. This, in turn, can result in coma and death. Many areas are doubling down on dreadful hordes of malaria, especially in Africa and India. These countries bear the burden of about eighty-five percent of the world's malaria cases. Malaria hits people with strained immune systems (e.g., pregnant women or people suffering from an unrelated infection) and toddlers the hardest.

Mosquitoes filled with malaria have strained society to the limit. But getting rid of mosquitos might be a bad idea. Fish use mosquito larvae as a food source, and bats and birds feast on the adults. Male mosquitoes even pollinate flowers in search of delicious nectar. So why do mosquitoes drink blood at all? Unfortunately, these pesky insects need proteins found in blood to lay their eggs. This is why male mosquitoes do not drink blood. Overall, the effort to eliminate mosquitoes is difficult, risky, and probably impossible. Malaria, with its severe symptoms and vicious attacks, will be sticking around for some time yet. Mosquitoes really do bug everyone!

Meet the Squad

There is actually more than one species of malaria, which gives the medical industry one whopping headache. All have home bases in the western Pacific, Africa, and southeast Asia. These nasties share the common name of *Plasmodium*. *Plasmodium falciparum* has headquarters in Africa. *P. falciparum* is also the deadliest species of malaria, as it clots the blood during its tenure in the bloodstream. *P. vivax* has mission control in Africa, but has outposts in Central America as well. This species of malaria isn't as bad as the notorious *P. falciparum*, but accounts for a good portion of malaria deaths. In the shadows are *P. malariae* and *P. ovale*—fortunately more like side characters in malaria's deadly gig. All four are the quintessential malaria representatives and are capable of severe damage. However, there is also a fifth species of malaria, *P. knowlesi*. This is the dominant malaria species in

Malaysia. *P. knowlesi* normally infects monkeys, but will hop to humans if given the chance.

Another trick up malaria's sleeve is drug resistance. *P. falciparum* in particular is extremely hard to treat; it is impervious to most commercial malaria-fighting drugs, or antimalarials. *P. falciparum* started to shrug off the classic malaria pill (chloroquine) in the late 1950s and early 1960s in South America and the Pacific Islands. *P. falciparum* is also resistant to an important group of antimalarials called artemisinins in Southeast Asia. Though multidrug-resistant *P. falciparum* is still rare globally, it is a looming threat we need to watch out for. *P. falciparum's* bad rap isn't a joke—this squiggly protozoa is a real pain in the neck!

Chloroquine is also being rendered useless by *P. falciparum's* "little brother," *P. vivax*. The first outbreak of chloroquine-resistant *P. vivax* was reported in 1989 near Papua New Guinea and Australia. Since then, these enhanced strains of *P. vivax* have been lurking in Ethiopia, Madagascar, and southeast Asia. However, this gang of *P. vivax* mostly skulks around down under in Australia. So if you are going to the Outback to see some kangaroos, it is best that you pack some malaria pills that are *not* made from chloroquine! *P. ovale, P. malariae,* and *P. knowlesi* can't be grown in a lab to be tested for drug resistance. However, there are some cases of these bad guys causing suspiciously severe infections.

Chloroquine resistance is a very big problem for some countries, because the alternatives are either too expensive or too toxic for practical use. And if you are also wondering why we haven't made progress on making new malaria drugs, the answer is surprisingly simple: it takes a long time for people to make new drugs, never mind "wonder drugs" that are effective enough to use. True, advances in medicine have made it easier for people to develop antimicrobial drugs. However, it still takes an average of ten years for a drug to be developed and ready for distribution around the world, not counting the amount of time devoted to finding the drug itself.

A Day in the Life

Part of malaria's success is that mosquitoes are found globally. A female mosquito can easily transmit malaria if it sucks blood from an infected victim. Malaria, in its slender sporozoite stage (think of malaria stages as Pokémon evolutions), then makes its way up to the mosquito's salivary glands, waiting for its next victim. When a mosquito bites a human, the wiggly sporozoites squirm through the bloodstream to the liver. Then, the sporozoites invade the liver cells by the thousands in a full-blown malarial onslaught. The sporozoites gradually become stouter and squatter as they calibrate themselves to invade the bloodstream once more. At this point, the infected liver cells aren't even called liver cells anymore. They are called schizonts, and, in fact, liver cells are not the only cells that become schizonts. Red blood cells infected by malaria are also called schizonts. (More on that later.)

Schizonts don't stay occupied for long. They eventually rupture and are demolished by swarms of fat sporozoites, now called merozoites. The merozoites then make their way back into the bloodstream. If their nasty work on the liver, which filters out the body's poisons and helps break down nutrients, isn't bad enough, the merozoites literally dig holes into the body's red blood cells. Inside the red blood cells, the merozoites transform into bad-tempered trophozoites, which look like eyeballs. As they grow inside the red blood cells, the trophozoites gradually change again until they resemble worms.

However, this relentless assault on the bloodstream doesn't continue indefinitely. Instead, some of the parasites become larger gametocytes, the forms that produce eggs, or oocysts, which resemble spotted blobs. These funky guys help produce more malaria parasites.

If another mosquito ingests the gametocytes, the gametocytes become orb-shaped macrogametes and microgametes, which look like balls of frayed knitting wool. Macro and microgametes combine to make an ookienite that digs into the mosquito's gut and transforms into an oocyst. Wiggly sporozoites burst out of the oocyst, and travel to the mosquito's salivary glands once more, waiting for their next victim.

Quick Henry, the Flit!

People have been trying to get rid of malaria in myriad ways. During the 1900s, the US and other countries sprayed entire fields and houses with a pesticide called DDT to get rid of the swarms of biting mosquitoes and other pesky insects. DDT does its dirty work by poisoning the bugs' neurons. It chopped down malaria cases in many developed countries. The pesticide worked so well that Dr. Seuss drew cartoon advertisements for a company that produced DDT-based pesticides called Flit. Seuss also made a household phrase ("Quick Henry, the Flit!") that showcased the chemical's potential as a bug-slaying maniac

With the help of DDT, malaria was eradicated in the US. However, the pesticide poisoned ecosystems on a global scale. Birds that ingested prey laced with DDT would lay eggs with brittle shells, breaking once their parents sat on them. DDT also decimated fish and amphibian populations. Humans even suffer from the effects of DDT: it runs the risk of causing cancer. Flies and other insects even developed DDT resistance, and became nuisances once more. The trouble with DDT is that it doesn't go away immediately. If DDT-contaminated water is slurped up by a thirsty prey animal, the pesticide lingers in the body, meaning that predators will often develop lethal concentrations of DDT after eating poisoned prey.

Perhaps to amend for his career drawing DDT cartoons, Seuss wrote the cautionary ecological tale *The Lorax*. A year later, the US banned DDT because of its devastating effects on the planet. However, some countries still use DDT to cope with mosquito-borne diseases. The war on DDT isn't over though; a chemical company in California dumped millions of barrels of DDT into the ocean. Still today, these barrels of DDT are on the ocean floor, like countless time bombs waiting to unleash an environmental catastrophe.

Delving through Panama

Keep in mind, though, that DDT wasn't all bad. Scientists estimate it saved the lives of 500 million humans from malaria. Before DDT was invented, mosquito-borne diseases were rampant in the tropics. Perhaps the most famous instance of this is when the Panama Canal was built. The French were the first people to attempt to build the canal. They faced hot, steamy jungles filled with massive swarms of biting mosquitoes, venomous snakes, and dangerous mudslides. If crossing paths with a ten-ton boulder cascading down a mountainside or a venomous viper weren't bad enough, malaria was even worse. Death rates by malaria skyrocketed in the jungles of Panama. According to Panama Canal builder Alfred Dottin, "The working conditions in those days were so horrible it would stagger your imagination. Death was our constant companion. I shall never forget the train loads of dead men being carted away daily, as if they were just so much lumber." To sum it up: it wasn't surprising that the French quit building the canal.

When Americans tried to build the canal, the mosquitoes attacked again. Health officials immediately took action. Yellow fever survivors William Crawford Gorgas, Joseph Augustin LePrince, and Samuel T. Darling led the charge. People installed screens on their gutters and windows to seal off pesky mosquitoes. Others drained pools down to the last drop or sprayed them with oil and mosquito-larvicides by the barrel. From hotels to mess halls, malaria-killing quinine pill-dispensers were found everywhere like gumball machines. People hired trained mosquito collectors to keep away the obnoxious insects. Officials even started a massive lawn mowing project to nix mosquitoes hiding in thick brush.

Quinine (another malaria drug) was especially crucial in the fight against malaria. It was originally used by the Quechua people of South America to treat diarrhea and the shivers. Found in the bark of the cinchona tree, quinine worked like a charm when it came to curing malaria. However, the drug was quite bitter, and patients had a hard time choking it down. The solution? Mix it with soda! Scottish doctor George Cleghorn created "tonic water" as a novel malaria cure by mixing this bitter medicine with a delightful soda. Tonic water is still a popular fountain drink but has a much lower quinine content.

By 1905, mosquito-borne diseases like yellow fever and malaria—once a major threat near the Panama Canal—were no longer rampant. Gorgas and his team were immortalized into medical history, such as in the names of universities and award titles. The Joseph Augustin LePrince Medal is awarded to people who have done an outstanding job in malariology, or the understanding and treatment of malaria.

Rolling the Dice

Coming back to modern times, malaria still hangs around and causes havoc in humans. A particularly nasty case of malaria befell a man named Stuart Ver Wys working in Haiti—a country in the tropical Caribbean Islands—while he was promoting human welfare. Wys slept in a tent near the edge of a city plagued by swarms of bothersome mosquitoes. The tent was not equipped with mosquito netting. Wys was not equipped with malaria pills. He did not get malaria on previous trips to the Caribbean and wasn't fond of the stomach trouble malaria pills gave him...so Wys rolled the dice.

A few days after he came back from Haiti, Wys came down with a fever and lost his appetite. As his condition worsened, he was brought to a local hospital, which misdiagnosed his malaria as the flu. Wys eventually got so exhausted that he started to lose track of his memory. He was then brought to urgent care at another hospital. Wys's breathing rate was so fast that his brain couldn't function right. His blood pressure was also at an all-time low. After doctors bombarded his veins with myriad drugs, he eventually recovered. However, Wys's recovery came at an immense medical fee of $23,383. Fortunately, Wys has learned his lesson and is now teaching about the dangers of malaria and how quickly and severely it can strike.

There are a couple of malaria vaccines in development as an alternative to the malaria pills Wys dreaded. One is called AS01 and is being distributed in sub-Saharan Africa and other malaria-prone regions. It is given four times to children under five months old—since protection wanes after a year—and will lower the massive annual death rate of 260,000 children in Africa. Widespread vaccine distribution is already happening in the African countries of Ghana, Kenya, and Malawi. AS01 is inexpensive and doesn't interfere with other antimalarials, making malaria prevention and treatment a much simpler business.

Taking to Our Heels

Since malaria has been bugging humans for quite a while, it is no surprise that we have developed defenses to combat it. Some people evolved sickle cell anemia, which causes their red blood cells to narrow into slender blades. Since malaria clots the blood when squeezing through the body's fine capillary blood vessels, these slender blood cells have an easier time sliding through these jammed spaces. However, sickle cell anemia can lower life expectancy and cause bouts of pain and fatigue. Slender blood cells can even clog blood vessels. Instead of staying and toughing it out, most people just packed a bag and moved away from areas infested with malaria. Malaria, along with a nasty cocktail of climate and drought factors, probably drove the first pioneering humans into the Middle East. Unfortunately, they took some of the pesky, malaria-infected mosquitoes with them!

Remember to Pack Bug Spray

To prevent malaria, always put on bug spray when traveling in mosquito countries. Even though bug spray smells bad, it is better than landing in a hospital. Regardless of whether you are sleeping in a five-star hotel or a flimsy old cabin, if there is a high risk of malaria or any other mosquito-borne diseases in an area, always use a bed net—think a cross between a mosquito screen and a butterfly net draped over your bed. In fact, people born in places prone to malaria have a significant survival advantage if they sleep under bed nets. Wear long-sleeved clothing if possible. Pack a standby, or a supply of malaria pills. However, shopping for malaria pills is a surprisingly tricky business. A man once used a so-called "malaria drug," which turned out to be a quack medical product that didn't even lower his fever. Turns out some substances that are claimed to prevent malaria are actually counterfeit. Try to probe out some information if there is chloroquine resistance in malaria in the area where you are staying. Drug resistance is a huge problem in the fight against malaria, but fortunately, there are alternatives to chloroquine. Be sure to book hotels that have good ventilation and air conditioning. Aside from helping to prevent malaria by blasting mosquitoes away, these creature comforts will also help you feel more at home. And when you are lounging by a beach in the tropics, you run the risk of contracting malaria. Don't slack off any malaria prevention guidelines while traveling!

It's Part of our History

Malaria is a deadly and ancient disease that has been closely intertwined with human history. We have been living under its iron fist for millennia, and finally, the long battle is in our favor. Despite recent setbacks, we have more diverse and effective anti-malarial technology than ever before. From bed nets to vaccines, we are finally gaining ground against malaria!

CHAPTER 8

DINO DISEASES

Y OU'VE PROBABLY BEEN OBSESSED WITH DINOSAURS AT SOME point in your life. Whether they're in Jurassic Park or featured in encyclopedias, dinosaurs are portrayed as invincible, multi-ton juggernauts of chomping teeth or long-necked titans stretching longer than an airplane. But while these kings of the past shoved down forests and duked it out with screams, roars, and bellows, these very same ancient lords were at the mercy of the microbes. Believe it or not, dinosaurs suffered from diseases. Badly. They had worms, contracted infections, and might have caught diseases common today, like malaria and tuberculosis. When you put yourself into the shoes or, in this case, claws of a dinosaur, this all makes sense: the swampy, hothouse kingdom of the terrible lizards was dripping with pathogens ready to do their dirty work. These feathery fiends and scaled scoundrels are showing their true colors now!

If you briefly leaf through this chapter, you may notice the diseases besieging dinosaurs all have something to do with bones. Dinosaur fossils usually don't preserve soft tissues like skin and organs, which decompose rapidly after dinosaurs die. Only bare bones are left behind, which gives us only a tiny piece of the massive puzzle of the world of dinosaurs. This little gimmick of nature fuels the misconception that dinosaurs were invulnerable, but the invincible dinos theory fails to hold up under closer scrutiny. In

reality, dinosaurs and other prehistoric critters probably suffered from a wide variety of infections. For now, here are a few discoveries scientists have found.

Busted Bones

A bulky, long-necked *Lufengosaurus* plods slowly around a forest clearing. Suddenly, it lifts its head and sniffs. Something big is coming. Something dangerous. It bellows, heaves itself upright on its hind legs, and begins to sprint for safety. As it makes a break for the safe haven of the forest, the dinosaur twists its head and sees its pursuer: a crested *Sinosaurus*—one of the big bruisers of early Jurassic China. Charging, the *Sinosaurus* scrapes two gashes into the *Lufengosaurus's* leg with a pair of vicious claws. As the *Sinosaurus* goes in for a bite, the *Lufengosaurus*—ignoring the pain shooting through its leg—kicks the *Sinosaurus* squarely in the side. With a sickening crunch, the *Sinosaurus* roars in anguish and withdraws, allowing the *Lufengosaurus* to limp away to safety.

Unfortunately, the *Sinosaurus's* bite is still lethal down the road. Several months later, some small dinosaurs and an entourage of chittering mammals all jockey for prime cuts of decaying meat on the dead dinosaur, which lies on a riverbank. But when the meat is gone, the bones are quickly enveloped in wet sand, soon becoming one of the bleached relics of the age of dinosaurs. Fossil analysis by pathologists some several hundred million years later shows that the *Lufengosaurus* had a nasty case of osteomyelitis. This offbeat-sounding name is medical speak for a bone infection.

This was bad news for our *Lufengosaurus*. The bacteria lingering in the bone caused countless problems. It made walking a chore, made the sauropod an easy target for other predators, and even allowed pathogens to access the brain! In the *Lufengosaurus's* case, the predator bite exposed the bone to harmful bacteria. However, another possible explanation for the busted bone is that pathogens spread from infected tissues and invaded the bone.

How did the scientists find out about this nasty infection in the first place? The busted bone was leaking abscess, which is the scientific term for pus. Even though you probably aren't a pale-ontologist, you could probably tell that something was definitely wrong if you looked at images of the bone. The dinosaur bone was splintered and it appears some sort of weird brown liquid (proba-bly fossilized pus) was flowing out of it.

Though we don't know what exactly caused the infection, pale-ontologists figured the culprit was most likely a pyogenic bacteria. Pyogenic bacteria, such as *Streptococcus pyogenes* (see Chapter 9), or *Staphylococcus aureus* (see Chapter 10), trigger the body to produce pus when they are attacking the body. This infection was very painful for the dinosaur, and probably contributed to its demise.

Another grisly case of osteomyelitis was in a *Titanosaur*. Just like *Lufengosaurus*, *Titanosaurs* were long-necked dinosaurs. However, they were much larger, and could reach the highest tree-tops in their ancient world. The infected *Titanosaur's* bones were covered with bumps and canals rushing with blood and pus. But this time, the infection was not the courtesy of bacteria, or even cancer. Scientists coated the fossil with gooey, protective resin and cut thin slices of the bone off to examine under the micro-scope. At a closer glance, the scientists discovered some surreal worm-like structures riddling the bone. Dubbed *Paleoleishmania*, this dust-mite sized critter was given the Halloween-worthy name of "blood worms."

Though *Paleoleishmania* is long gone, its possible modern-day cousin *Leishmania* is a human parasite spread by sand flies that causes similarly gruesome-looking lumps on its victims today. It is a good thing paleontologists only had bones to look at, for in life, this *Titanosaur* looked truly horrifying. The bony protrusions made nasty, bleeding bumps protrude out of the flesh of the dinosaur, making the otherwise innocent-looking dinosaur something out of a nightmare!

Dolly the Diplodocus

A *Diplodocus* dubbed "Dolly" by scientists had a life that all long-necked dinosaurs should have. She munched away at trees, plodded across floodplains, and trampled smaller dinosaurs underfoot. But when she was in her teens, Dolly fell ill. She often coughed and spewed mucus (think big gobs of it falling from the sky), lost weight, and eventually died. What had happened? *Diplodocuses* usually have long and peaceful lifespans, with some reaching the ripe old age of eighty. What type of infection could kill such a big and mighty dinosaur like Dolly?

Paleontologists examining Dolly's bones soon found out. Only the skull and seven vertebrae—or neck bones—remained, making it easy to sniff out Dolly's infection. While they were alive, *Diplodocuses* were equipped with a huge amount of air sacs to lighten their bodies, just like birds. So despite their size, they were "only" thirteen tons, not much heavier than a killer whale. Despite putting them on the skinny side of long-necked dinosaurs, these air sacs made Dolly and her long-necked kin prone to infection. The team of "dino doctors" discovered bunches of bizarre, broccoli-shaped protrusions in the air sac sockets of the bones. Similar bony knobs are the hallmark aspergillosis infections (See Chapter 5 for more info on this naughty fungus) in modern-day birds. Since dinosaurs are technically birds, it is very possible—if not probable—that they suffered similar infections.

Scientists can't help feeling sorry for the pneumonia-stricken dinosaur. *Aspergillus* infections are a pain in the neck, with balls of dead fungus, mucus, and scabs (called aspergillomas) accumulating in the respiratory system. As one of Dolly's "doctors" put it, "We've all experienced these same symptoms—coughing, trouble breathing, and a fever—and here's a 150-million-year-old dinosaur that likely felt as miserable as we all do when we're sick."

Dolly was likely immunocompromised too. Healthy birds exposed to *Aspergillus* spores don't usually develop an infection.

But Dolly also could have been exposed to a large amount of spores, making the chance of infection high. So there are two hypotheses on how Dolly was infected: Dolly was immunocompromised and fell victim to an *Aspergillus* infection, or there were a particularly large number of spores in the area she lumbered around in. Regardless, Dolly's story is a sad one. Poor girl.

The Hanger Games

Unlike Dolly or the *Lufengosaurus* mentioned above, you might've seen Sue the *T. rex* before, especially if you've visited the Chicago Field Museum, where the titanic skeleton is on display. If you look carefully, you might see bunches of holes in Sue's lower jaw. Most people who see these peculiar pockmarks claim they were the result of a clash of titans, or maybe the work of smaller dinosaurs gnawing at Sue's jaw after her death. But the strange holes were not gouged out by a rival *T. rex* or even other scavengers; the rest of the bones are relatively intact.

This led paleodoctors flocking to the scene again. Careful analysis of the bones showed that, unlike Dolly, Sue was twenty-eight (that's retirement age for a *T. rex*) when she died. Growth rings in her bones were helpful for diagnosing the *T. rex's* age. But this analysis also uncovered a nasty parasitic bone infection called *Trichomoniasis* in her bones. Some birds, like pigeons, contract *Trichomoniasis* but shrug it off quite easily. However, a few unlucky birds such as falcons and hawks are affected by this protozoa, which nibbles away at the jawbone and creates holes. The immune system tries to fight back, but creates nasty-looking lesions...and gives these birds unbelievably "fowl" breath. Sue's immune response to her infection was almost exactly the same as a modern-day bird!

Where did Sue pick up this infection? Probably from one of her various dinosaurian steaks. Sue might've caught the parasitic infection after catching a duck-billed *Edmontosaurus* (a *T. rex's* favorite meal), or a wayward *Triceratops* that was unlucky enough to be thrashed to pieces by her jaws. *Trichomonas* relied on a predator-prey cycle in dinosaurs, just like many other parasites past and present. Though ugly, *Trichomonas* shows how evolution shapes parasites and other diseases through time.

Sue probably starved to death because of this complication. *Trichomonas* makes eating or drinking extremely painful for birds, so it is reasonable to assume it made dinosaurs like *T. rex* hangry—but unable to eat—as well. And now you somehow feel sorry for *T. rex*.

Sneezing since 245 Million Years Ago

As the sun peeked over the horizon some 245 million years ago, a strange marine reptile, resembling the infamous Loch Ness Monster, named *Cymatosaurus* hauled itself out of the sea. As turquoise waves laced with creamy white foam lapped away at the beach, the *Cymatosaurus* plopped down in a comfortable spot in the sand. For this seafaring lizard, the new day seemed perfect. But what the *Cymatosaurus* didn't know was that a lethal disease was slowly creeping through his body, tainting his organs with an illness that his immune system was hard pressed to fight off. The sickness attacked his lungs and even crawled inside the reptile's bones, making swimming a pain. Eventually, the *Cymatosaurus* succumbed to this pathogen. But who was the guilty microbe?

When examining the *Cymatosaurus's* fossils, scientists found swaths of bony protrusions in the reptile's ribcage. Scientists weren't sure if they were tumors, fractures from sea battles, fungal infections, or scurvy (a disease caused by a lack of vitamin C). However, closer examination revealed the bony nodules looked astoundingly similar to lesions created by tuberculosis (known as TB for short). Before that, people thought that TB emerged relatively recently, from 10,000 to 100,000 years ago. Apparently, its ancestors have been bouncing around in the lungs of animals for hundreds of millions of years, shocking the scientific community.

There is also additional evidence to back up the fact that the *Cymatosaurus* suffered from TB infections. TB can seep into the bone and cause tremendous damage. Plus, *Cymatosaurs* are the reptilian counterpart to modern-day seals. Seals are particularly susceptible to TB, so it's reasonable to assume similar reptiles, such as *Cymatosaurus*, suffered the same.

In-flight Turbulence

Pterosaurs are fuzzy, beaked creatures with leathery wings commonly called "pterodactyls" by the general public. Despite being absolutely adorable, ancient predators still attacked them out of hunger (No respect, huh?). Some pterosaurs such as *Pteranodon*—a massive beast twice the size of the largest bird today—flew out to sea to catch fish to fill their stomachs, so it is no surprise they encountered sharks hungry for a furry, flying snack. A shark tooth was even found wedged in an unlucky pterosaur's neck! Even if these flying reptiles survived the injury, the pterosaur would find flying difficult.

Pteranodon fossils dug up by paleontologists have shown a bonanza of bony bumps peppering their jaws and wings. Experts have diagnosed this as necrosis, or tissue death/decay. This condition can spring up from a variety of causes: disease, injury, and even radiation exposure (one of the many reasons why Godzilla wouldn't exist). Back in the day, the pterosaurs were either sick or attacked by predators and developed these uncanny bony structures.

But the bumpy bones hide what is actually going on. Though the pterosaurs' bones didn't get fractured or broken, necrosis surely was painful. Necrosis mostly affects flesh. When tissue affected by necrosis dies, it leaves behind a disgusting, leathery, brownish-black paste of dead cells called a necrose. These necroses are very prone to infection. Smart as they were back in the day, pterosaurs had not invented rubbing alcohol. If necrosis is allowed to progress, a serious infection called gangrene, which is caused by oxygen-hating bacteria, develops. It's a miracle that the pterosaurs suffering from necrosis were even fossilized!

Swamp Suckers

A *Deinocherius* waded across a swamp in Mongolia eighty million years ago, seeking refuge from merciless swarms of biting midges. Its thick, gray feathers and massive, humped back were caked in swamp muck as it plodded onward. The atrocious-smelling mixture attracted more of the pesky insects. Finally, the *Deinocheirus* waddled clumsily out of the swamp and into a forest, which was free of the swarms of insects. However, the dinosaur's troubles had just begun because the biting midges carried a secret, microscopic weapon: an ancient form of malaria. The *Deinocheirus* rapidly grew weak, barely able to stand on its shaking legs. Just when the disease was at a fever pitch, a *Tarbosaurus* bounded into the clearing where the *Deinocherius* lay. Roaring with triumph, it lunged. Before the *Deinocheirus* could wheel around and jab at the *Tarbosaurus* with its sharp claws, the *Tarbosaurus* restrained the bucking dinosaur with its short yet stocky arms as it went in for a bite. It wasn't long before the *Tarbosaurus* enjoyed a saurian steak.

Initially, malaria was thought to be exclusive to more recent times. The first fossils of the parasite (trapped in insects, of course) were dated to about twenty to fifteen million years ago in the Dominican Republic. However, some spectacular finds show a 100-million-year-old biting midge preserved in amber carrying a "grandfather" of malaria. The fossil was discovered by George Poniar Jr., a researcher at the College of Science of Oregon State University. This means malaria ravaged the globe eighty million years earlier than expected...and woe to even the largest dinosaur that contracted the disease.

Malaria is very severe in birds. In fact, it is so severe that several species of birds have been driven to the brink of extinction by this parasitic scourge, especially on tiny islands. One famous bird almost done in by this parasite is the Hawaiian crow. The introduction of pigs and rats killed many crow chicks, but malaria was the real killer. It was nearly wiped out by this protozoan, and repeated outbreaks only made matters worse. To this day, the Hawaiian crow is still very much endangered, though it is protected by law and buffered by zoo-based breeding programs.

Malaria develops in birds or mammals, then hitches a ride on insects to breed. It is possible the biting midges of the Cretaceous swarmed the skies in huge numbers, carrying their lethal cargo. Dinosaurs are the ancestors of birds, so it is probable they suffered from the same disastrous effects. In fact, some people even suggest ancient malaria contributed to the extinction of the dinosaurs! However, dinosaurs persisted for at least thirty-five million years more scratching at these pesky midges, so it isn't likely malaria ended the reign of the dinos, though it could've struck the final blow. We should be thankful that the same fate doesn't apply to us. Malaria claims twenty-four million lives each year!

Did Dinosaurs Get COVID?

Could dinosaurs have gotten coronavirus like COVID-19? Birds today suffer from a wide variety of respiratory infections, but not COVID-19. Because dinos possess a lung system similar to birds, these reptiles don't seem like promising hosts for SARS-CoV-2. Scientists are building a huge database of duck-billed, plant-eating dinosaurs to see which ones were most susceptible to diseases like coronaviruses. However, birds and reptiles are not seriously affected by these viruses. SARS-CoV-2 (or one of its predecessors) would've more likely infected our mousy mammal ancestors living in the trees or groping around in burrows.

Microbes Rule

Though dinos seemed like mighty titans that ruled the landscape, they were actually at the mercy of an even more powerful conqueror: the pathogens. They were here long before the dinos and will be here long after we're gone. They will always be hiding in the shadows but will always rule our planet. Watch out for your crown, *T. rex!*

CHAPTER 9

THE TWO-FACED BACTERIUM

WHAT LOOKS LIKE A CATERPILLAR AND IS THOUSANDS OF TIMES smaller than a grain of rice? *Streptococcus*, the two-faced bacteria. There are many species of streptococci, which is plural for *Streptococcus*. Some cause pneumonia, cause nasty throat infections, and even gnaw on flesh, while others whip up platters of cheese and bowls of creamy yogurt. Whatever it does, a bacterium doesn't need to be completely "good" or "evil," it could draw the line somewhere in between.

There are many "groups" of streptococci, but two groups stand out: Group A (GAS for short) and Group B (GBS for short). Group A Strep—aka *Streptococcus pyogenes*—cause strep throat, nasty skin infections called impetigo, and kidney infections. Group B—aka *Streptococcus agalactiae*—kindle brain infections, or meningitis in newborn babies. Other streptococci cause pneumonia, bone infections, bloodstream infections, and even prompt the body's immune system to go into overload in a lethal process called septic shock. However, these bacteria share some commonalities. All streptococci are anaerobes and love rummaging around in low-oxygen environments, though most can also grow in air. Many of them exist harmlessly in the mouth, on skin, inside the nose, and/or intestine. And all are nonmotile, or unable to move, contrary to the drawings of this book. My cartoon streptococci wriggle around for artistic licenses only!

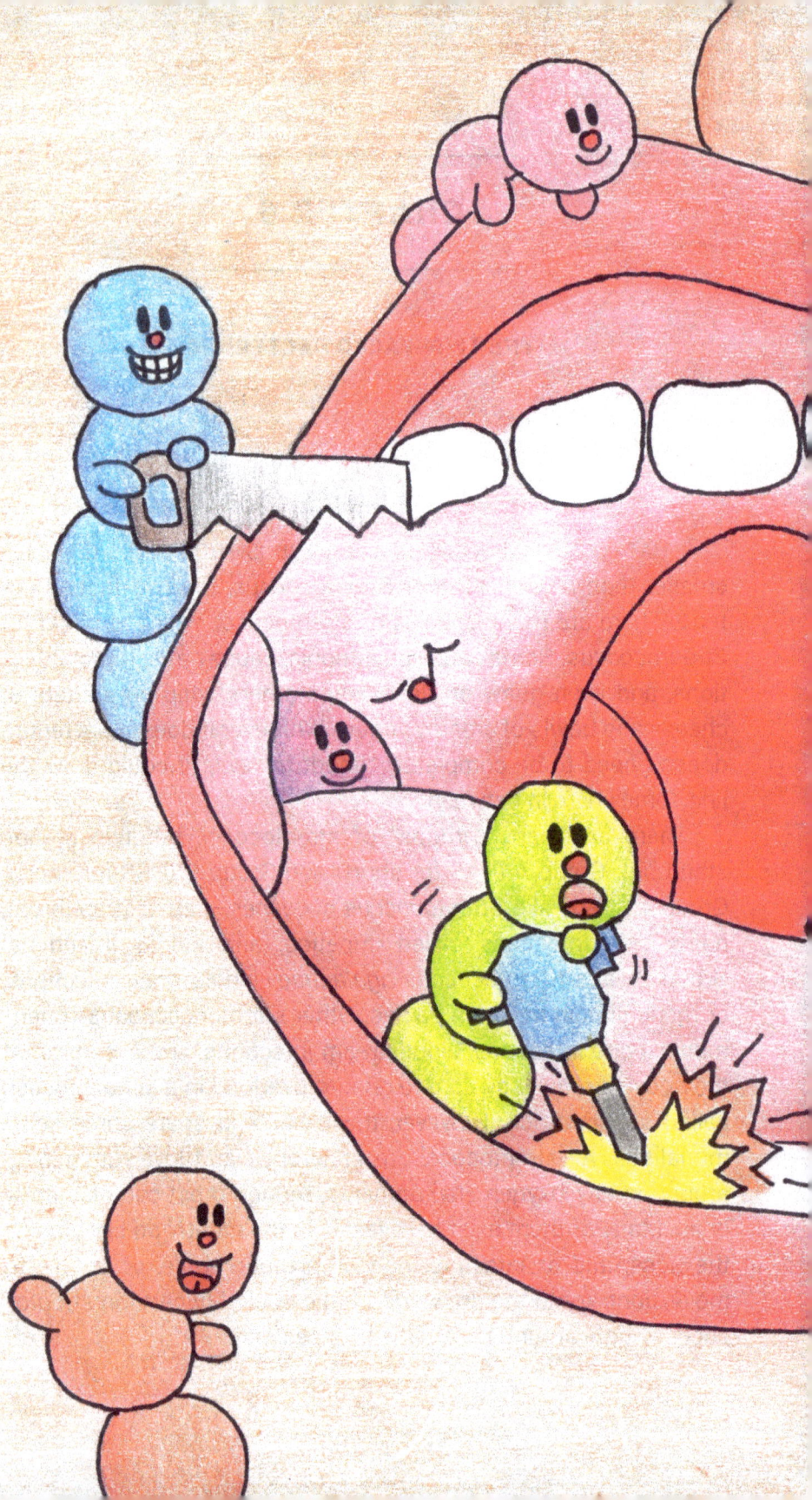

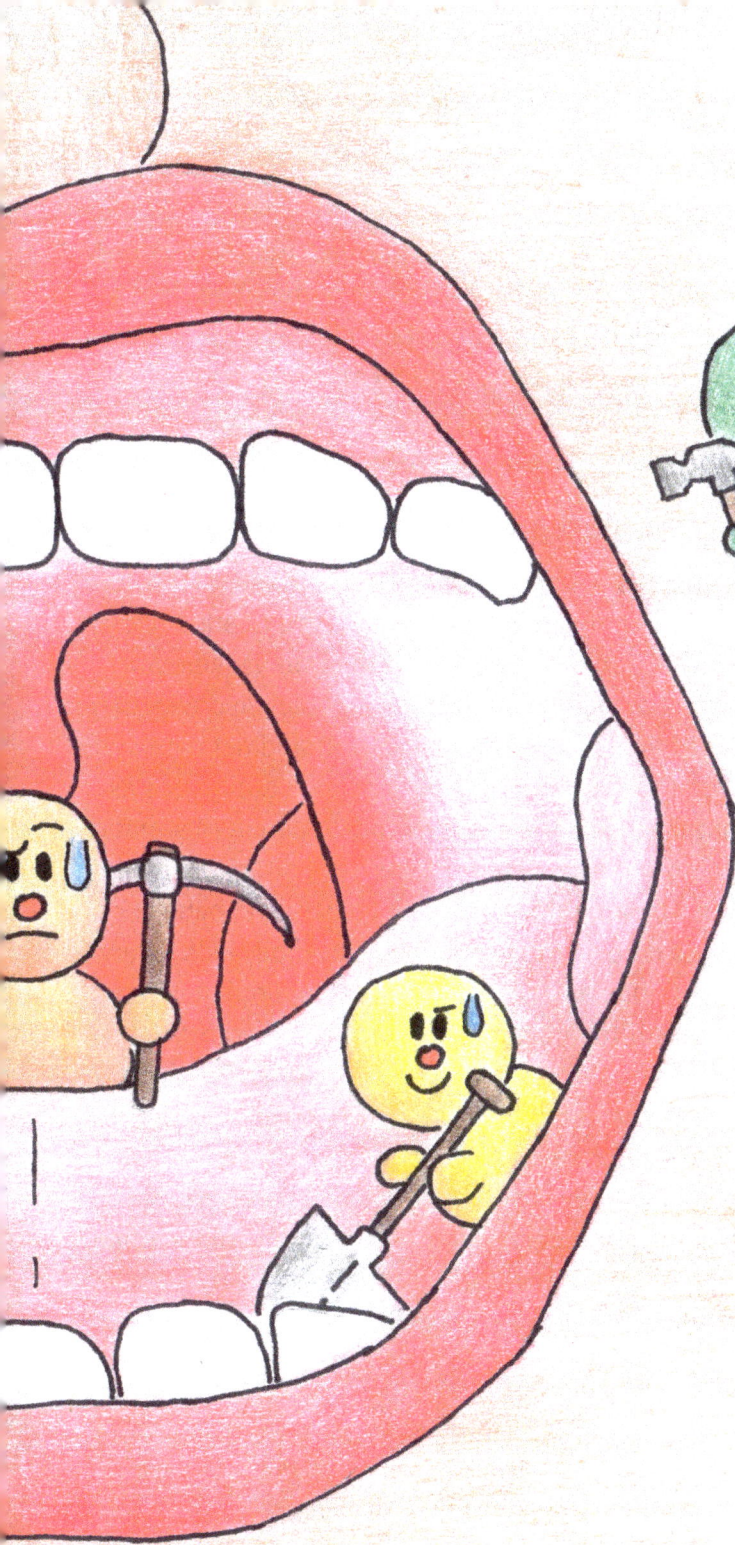

Throat Blues

Many people across the world find themselves sweating in bed with a fever and an excruciatingly painful sore throat each year. They haven't just caught an ordinary cold: they are the victims of GAS, or *S. pyogenes* (say pie-oh-geneez). They are ravaged by strep throat. Oftentimes, their upper jaw is peppered with red dots. When these unfortunate souls swallow, it feels like they have swallowed razor blades. They suffer from unsettling rashes, bouts of nausea, and throb from nasty fevers.

S. pyogenes hitches a ride on sneezes and sneaks into the mouth and nose, making this mischievous microbe nearly impossible to avoid. To make matters worse, there is no vaccine for *S. pyogenes*. The best that victims of this microscopic miscreant can do is to drink lots of fluids (yes, Slurpees count) and get plenty of ZZZ's. It's important to conk out *S. pyogenes* with antibiotics like penicillin, or it might ravage the kidneys and/or cause rheumatic fever, an inflammatory disease that affects vital organs, such as the heart. To top it off, strep throat causes anxiety and behavioral problems in mice, so it could make humans, especially kids, jittery.

Inside the body, things are tense during a strep throat infection. White blood cells are often portrayed as simple killing machines releasing antibodies or mindlessly gobbling up germs. But there is a lot of subtlety to these micro warriors. When duking it out with streptococci, white cells release a special blast of amino acids called antimicrobial peptides. These nifty chemicals burn holes in *S. pyogenes's* cell wall and serve as a battle cry for more white blood cells to rush to the scene. With a little help from antibiotics, the battle against *S. pyogenes* is easily won. However, not all microbial wars are so easily won.

Om Nom Nom

If carnivorous plants give you a bad scare, do you dare to go toe to toe with carnivorous streptococci? The scientific name for flesh-eating streptococci, or any other flesh-eating bacteria in general, is known as necrotizing fasciitis (say neck-kruh-tai-zuhng-fa-sh i-ai-tuhs). Necrotizing fasciitis is an extremely serious illness. It can progress to lethal levels in a matter of hours. The only cure for the disease is a convoluted concoction of powerful antibiotics. There are two types of necrotizing fasciitis: monomicrobial and polymicrobial. Monomicrobial necrotizing fasciitis (I know, lots of long words today) is caused by GAS but can also be induced by *Staph aureus* (if you wanna know more about *S. aureus*, flip to Chapter 10), while polymicrobial necrotizing fasciitis is the courtesy of a devastating dream team of lethal microbes. If you start to get the shivers, move on to the next section, "Cheesy Business."

Necrotizing fasciitis is extremely barbaric, sort of like a micro version of Genghis Khan or Attila the Hun. The disease causes skin to rapidly swell and turn red, tender, and warm. Most doctors think this isn't a big deal. At a first glance, these early infections just look like another bruise you get after falling on the pavement—why worry? But soon, sickly brownish lumps appear on the ankles as streptococci bust open red blood cells, creating a mishmash of scabby connective tissue. This, known as brawny edema, begins occurring two days after infection. As swarm after swarm of vicious microbes batter infected tissue, giant blisters oozing with pus and blood called bullae form. Tissues wither to black and rot (known as gangrene) as the infection carves its way through the body.

If someone is infected with flesh-eating streptococci, there is a thirty-five to sixty percent chance of them dying a grisly death. To treat necrotizing fasciitis, doctors amputate infected limbs and treat patients with a vicious cocktail of antibiotics and antibodies. Early treatment is crucial. However, this is difficult since the early stages of necrotizing fasciitis resemble normal wound infections. Everyone can be hit hard by flesh-eating streptococci, but the elderly and infants are the most vulnerable. To save your buns from flesh-eating bacteria, treat GAS infections as early as possible. Flesh eating bacteria can be spread through insect bites, so spritz some bug spray on yourself when rambling around in the great outdoors. Properly care for your wounds with a wash, a dollop of antibiotic cream, and some bandages. Fortunately, necrotizing fasciitis is a very rare infection. We don't want microbes eating us...after all, we are the ones that are supposed to be gobbling up them!

Cheesy Business

Harmless microbes are normally ingested in large amounts whenever you chow down on the goodies in your fridge. However, some microbes are purposely added to foods. Certain types of streptococci are the secret ingredient in cheese. No, cheesemakers don't dump barrels of nasty bacterial sludge into our cheese. The micro cheese connoisseur in question is called *Streptococcus thermophilus*. *S. thermophilus* is a harmless little fellow that gives Swiss cheese its tangy flavor.

S. thermophilus is often bestowed with the prestigious title of "probiotic." But what are probiotics, anyway? Probiotics are *live* microorganisms that are believed to provide health benefits when consumed. Products containing probiotics (e.g., yogurt, cheese, and kimchi) are worshiped as healthy foods. People claim probiotics crowd out more harmful bacteria, boost metabolism, and eliminate toxins. However, studies have shown probiotics do not provide any benefits to the body. Probiotics are like the new kids on the block—they are kicked around by your own microbiome before being disgracefully ejected out of your digestive tract.

Fortunately, probiotics can find refuge in cheesemaking and avoid this fate. *S. thermophilus* rustles up cheese with two of its friends: *Lactobacillus*, a probiotic added to food to beef up the intestine's good bacteria; and *Cutibacterium*, a not-so-cute pro-biotic that creates zits and pimples outside of its culinary career. *S. thermophilus* and *Lactobacillus* process the cheese and pro-duce lactic acid. The *Cutibacterium* then fuels up on lactic acid and releases a delicious combo of acetate, propionic acid, and carbon dioxide. Acetate and propionic acid give Swiss its delicious taste, while carbon dioxide bubbles give the cheese its charac-teristic holes. This process is called fermenting. The more time bacteria are allowed to ferment cheese, the more holes pop up. The more holes, the stronger the flavor. However, too many holes in one hunk of cheese can, predictably, make it collapse, crumble, or even explode!

Don't like Swiss cheese? *S. thermophilus* also helps make your strawberry yogurt. *S. thermophilus* is best buddies with *Lactobacillus bulgaricus*. The two pals give yogurt its squishy tex-ture. *S. thermophilus* and *L. bulgaricus* ferments lactose in milk into lactic acid, which is why yogurt tastes sour. This lactic acid causes a protein found in milk called casein to form heaps and glop together into yogurt. Yum.

However, it's too bad the next species of *Streptococcus* in this chapter isn't such a great chef.

Swimming Pools in the Lungs

You might hate going to the doctor and seeing that dreaded vaccine needle. However, shots aren't just protecting you against old-timey diseases; they also protect you against a much more common and deadly disease known as *Streptococcus pneumoniae*. This little critter is nasty, and unlike its happy-go-lucky cousin *S. thermophilus*, *S. pneumoniae* rampages through the lungs and causes 150,000 people to land in the hospital each year.

This clever microbe has a few tricks up its sleeve. *S. pneumoniae* bacteria can "talk" to each other using special chemicals called autoinducers. This is known as quorum sensing. Unlike your small talk about the latest Xbox games or the new kid next door, *S. pneumoniae* are no-nonsense. Their conversations are all business. When there are too many microbes in one area or when nutrients are low, some *S. pneumoniae* bacteria will be convinced to poison and kill themselves for the greater good of the community. Sounds awfully harsh to me. However, quorum sensing comes with a big perk: *S. pneumoniae* can talk to other microbes. No Google Translate needed!

S. pneumoniae are also wolves in sheep's clothing. The bacteria can conceal themselves under a blanket of jelly, making it hard for immune cells to detect them. This microbial disguise makes them look innocent as flowers! When a macrophage (the champion germ-eating white blood cell) *does* unveil *S. pneumoniae's* ruse, the bacteria is spat out by the macrophage thanks to its protective shield of goop. Sort of like when you find a glass of "Coca-Cola" on the counter, eagerly chug it down, then find out it's bitter coffee instead. Stealthy!

When *S. pneumoniae* jukes the immune system and infiltrates the lungs, it begins its dirty work. *S. pneumoniae* infects the alveoli sacs—air sacs in the lungs that transport oxygen into the blood. Immune cells pour into the lungs in an attempt to fight off the bacteria, causing pools of fluid to seep into the alveoli. While pneumonia is not always lethal, severe cases are equivalent to being drowned by microbes from the inside. As more "swimming pools in the lungs" are built, an infected person will struggle to breathe until he or she will need a lung machine to do the work for the fluid-laden lungs.

If it gets the chance, *S. pneumoniae* can spread to the blood-stream and cause sepsis or march up to the brain and cause meningitis. The bacteria also cause an array of minor infections, such as ear problems. Symptoms of this pesky little bugger include joint pain, fever, chest pain, a stiff neck, chills, fits of coughing, insomnia, and difficulty breathing. Best to activate full defense mode and scamper over to the doctor when this microbial rascal is around!

There's Always a Plan B

Group B Strep, or GBS for short, is relatively harmless compared to its grouchy cousins. Most types of GBS lounge about the human body for a few days, then pack their suitcases and leave without a problem. However, this book features infectious diseases, so—you guessed it—GBS infections are commonplace across the world. Anyone can get a severe GBS infection, but babies are particularly vulnerable. Infants may contract GBS from their mothers when they're born. In fact, GBS is a major cause of meningitis and blood-stream infections during an infant's first three weeks of life. GBS infections come in two varieties, neither of them good: early-onset, which occurs during a baby's first week of life; or late-onset, which occurs during a baby's first to third week of life. Both cause babies to be irritable, fatigued, struggle to breathe and eat, and turn blue.

However, GBS infections in babies are quite rare. A newborn has a one in 200 chance of contracting GBS if their mother hasn't taken antibiotics to kill the bacteria, and a one in 4,000 chance if their mother has. Among the three million babies born in the US each year, there are only 930 babies that develop early-onset

GBS diseases and 1,050 babies that contract late-onset GBS diseases. But when GBS does strike a newborn, the stakes are high. Approximately seventy percent of babies infected with GBS die. Children and adults have a much safer gamble in the GBS casino; they have a one in 10,000 chance of contracting a severe GBS infection. However, the elderly have a one in 400 chance of developing the infection when exposed to the bacteria. But we shouldn't fret. Usually, GBS is as tame as a bunny.

One big catch? There is no vaccine for this two-timing microbe. It is still being developed.

Get Your Scruffy Face Out of Here!

There is a large assortment of treatments devoted to knocking out streptococci. The first streptococci vaccine was made by Sir Almorth Edward Wright, who also synthesized tuberculosis and *Salmonella* vaccines. He made the first *S. pneumoniae* vaccine by zapping the bacteria under extreme temperatures and then used the dead microbes as a vaccine. The slain bacteria weren't able to attack the body, but could still evoke an immune response. This proto-vaccine was tested on gold miners in South Africa and it worked! Sir Wright also tested his life-saving vaccines in Papua New Guinea. The vaccine was improved by many brilliant scientists, including Franz Neufeld, Alphonse Dochez, and Oswald Avery. Unfortunately, the vaccines were forgotten during the 1950s when penicillin became commercially available. Vaccines began being manufactured in 1960 when researchers discovered people still had a twenty-five percent chance of dying from *S. pneumoniae* even with the help of penicillin. A revamped vaccine was made in 1977. Like Sir Wright's vaccine, it was used on trial in gold miners in South Africa. The newer vaccine was later touched up in 1983.

Penicillin is also a champion streptococci smoosher. However, one third of all streptococci can shrug off penicillin. Fortunately, there are backup antibiotics for streptococci, such as vancomycin.

But time is running out. We must stop the rising tide of antimicrobial resistance. No treatment is foolproof, but we can always make it better. It's a big mistake to stop creating and say our current arsenal of antibiotics and vaccines is good enough.

Streptococcus is a very versatile bacterium. It can put together scrumptious platters of food, yet cause flesh to rot in the blink of an eye. It lives in the air, the soil, the gut, and myriad other places. One moment, it can be a benevolent angel. Another, it can be a fearsome devil. The power of streptococci can be harnessed to our advantage, but cannot be underestimated. We need to keep our eyes wide open for this two-faced bacterium.

CHAPTER 10

CATTYWAMPUS CUSTARD

A CUSTARD PIE IS SET ON THE TABLE RIGHT AFTER YOU RETURN from a hardcore day at school. Such a perfect golden crust. So wispy and light. So delectably scrumptious. If you are a custard pie fan, you aren't alone. A type of bacteria loves custard pies even more than you do. If stored improperly, cream pies, pastries, or any other food in general can host an orb-shaped bacteria called *Staphylococcus aureus*, also dubbed "golden staph" (more on that later). When the bacteria aren't gobbling down literal upside-down cake, they live in your nose, on your skin, and even in your bathtub. While usually harmless, *S. aureus* produces a nasty toxin called enterotoxin if it slinks into a cut or your digestive tract. Enterotoxin induces diarrhea, vomiting, dehydration, and can even change a person's heart rate in a matter of hours. But like a toddler's tantrums, *S. aureus* infections come as fast as they go. Symptoms can disappear in a few hours.

Though it might be famous for stirring up stomach trouble, *S. aureus* can cause a variety of other problems. With its ability to creep into wounds, *S. aureus* is known as "golden staph" because it triggers the body to ooze a sickly golden pus, and the color of the bacteria once grown in the lab are also golden. But this is just the beginning of the long list of havoc these bacteria can cause. Most people who contract *S. aureus* don't get treatment in time, and unfortunately, some are not strong enough to ward off the

infection. Things can get especially gruesome if S. aureus invades the bloodstream and causes sepsis. Sepsis is not an infection, but the immune system's lethal counterattack when the body overreacts while fighting blood-borne bacteria. During sepsis, the body releases an overwhelming amount of chemicals and attacks itself; this chemical avalanche can lead to inflammation, organ damage, vomiting, diarrhea, and extreme agony.

In rare cases, some particularly ghastly S. aureus bacteria can even eat away at swaths of flesh. This is called necrotizing fasciitis, which is also known as the less tongue-tying but more horrendous name of "flesh-eating bacteria." If these ruthless zombie-bacteria start to chip away at muscle tissue, this is known as necrotizing myositis. These dangerous bacterial infections don't actually munch away at your innards, but do cause mass tissue death. Not a very preferable alternative! Regardless, both types of destructively hangry S. aureus are lethal if they are not treated immediately.

Flesh-eating aside, S. aureus can also whip up pneumonia by causing the lungs to fill up with mucus and fluid. This can lead to sepsis, organ failure (shock), and even necrotizing pneumonia. S. aureus is also the largest cause of surgical infections in the world. This sly bacterium even outranks the infamous Pseudomonas aeruginosa in its ability to slip into surgical incisions. These infections are the result of inadequate surgical practice. Treating S. aureus surgical infections is complicated; it involves sterilizing the infected area, removing dead tissue, draining pus from the infected area, and to top it off, injecting antibiotics into the veins. Ouch.

One Tough Cookie

S. aureus is a tough cookie as well as a nuisance. It can survive in dusty dry environments and lie dormant for months. The bacteria can also survive in both aerobic (oxygen-rich) and anaerobic (oxygen-poor) environments. S. aureus also has a bunch of tricks up its sleeve to combat drugs. A spray of enzymes such as "penicillinase" or the wackier "aminoglycoside-modification enzymes" often does the trick. Despite their quirky names that make them sound like madcap concoctions, these enzymes have been honed by the blades of evolution and viciously break down antibiotics. *S. aureus* can also unleash a bonanza of proteins to bind to and neutralize antibiotics. If these salvos of proteins and enzymes miss their mark, efflux pumps rapidly suck up and remove any trace of antibiotics from *S. aureus's* innards.

Equipped with an arsenal of its powerful defenses, *S. aureus* seems like the ultimate living fortress; however, *S. aureus* is also skilled in offense. With a bit of cunning manipulation, this bacterium can even make the body's own immune system attack itself. Neutrophils, the most common white blood cell in the body, produce NETs (Neutrophil Extracellular Traps) that trap and ensnare bacteria. However, *S. aureus* can break down NETs into a chemical called 2'-dAdo—short for the more kooky-sounding 2'-deoxyadenosine—with an assortment of special enzymes. dAdo is toxic to macrophages, the big bosses of the immune system. Because of this chemical warfare, macrophages are often absent in staph-infected wounds, allowing *S. aureus* to have full dominion over an infection. However, studies have shown that if the production of dAdo is shut off, macrophages can return. Though *S. aureus* is an excruciatingly irritating microbe, it is hard not to admire this little fella's trickery.

Knocking Out the Champs

Besides skulking about in those custard pies, *S. aureus* also lurks around objects that come into close contact with skin, such as towels, shaving razors, toothbrushes, or just plain old soil. If these contaminated items bump into a wound or cut, *S. aureus* quickly marches into the body. *S. aureus* is also blasted out by sneezes or coughs. But the bacteria's true power is unleashed when it forms biofilm. *S. aureus's* biofilms are resistant to a wide variety of chemical attacks, and can be made from a medley of substances. Sticky sugars called polysaccharides, a protein goop named amyloid, fibrin (the stuff scabs are made of), and even DNA are all used. *S. aureus's* biofilm makes it extremely dangerous. If it infects the heart, it can cause a potentially fatal inflammation of the heart's inner lining. This is known as endocarditis. This lining, or endocardium, is one of the muscles that allow your heart to beat. Worse, *S. aureus's* biofilm can even glue the heart's valves together. This gooey bacterium can also cause infections inside the eyeball (endophthalmitis). For newborn babies, especially premature babies, *S. aureus* is extremely dangerous. Wash your hands thoroughly before you cuddle a baby!

S. aureus can also topple the most jacked up of people as well. People who practice boxing, karate, and other mixed martial arts (MMA) are particularly vulnerable to *S. aureus* infections. They often develop injuries that allow *S. aureus* to squeeze into the body. Boxing champ Robert Whittaker developed an *S. aureus* infection in his stomach. This wasn't any normal stomach bug; it was a full-on invasion! The infection became so severe that his organs began shutting down. After one arduous year, he finally recovered. Whittaker is still duking it out in the ring today. Pro wrestler Kevin Randleman also contracted a severe *S. aureus* infection. The infection shut down his kidneys and liver, created a gaping hole in his arm, and threw him into a coma. Randleman barely made it out alive and later died of pneumonia when he was forty-four.

It isn't just the champion brawlers who are at risk of these brutal staph infections: anyone who practices MMA is vulnerable. But if you are proud of your black belt, you can still practice martial arts. However, make sure to attend to wounds with antibiotic cream or rubbing alcohol. If the wound becomes infected, see a doctor for advice.

One *Staph* to Rule Them All: MRSA

As bad as they may be, the dangers of MMA-related staph infections are dwarfed by MRSA, or **M**ethicillin **R**esistant *Staphylococcus*

Aureus. MRSA can shrug off a whole laundry list of antibiotics. This mutant *S. aureus* emerged in the 1950s, when antibiotics like methicillin and penicillin were abused and overused. Even before penicillin was widely used, some *Staph* strains already possessed resistance against this "miracle drug." MRSA is a tough nut to crack; it can even hide in the cartridges of street drug injections. People taking street drugs frequently share dirty needles, allowing MRSA to burrow into their bodies. The problem here is crystal methamphetamine, or crystal meth. This lethal drug creates an appalling itch under the skin. With each scratch, tiny wounds are opened, increasing the chances of MRSA entering. MRSA is one of the countless reasons why illegal drugs are such a big problem!

MRSA also thrives inside bedbugs and other biting insects. These pests serve as vectors—the microbe equivalent of public transportation—for MRSA, allowing bacteria to bounce around from person to person. There are many species of bedbugs crawling around on bats, birds, cats, dogs, rodents, or anything else warm and fuzzy. These critters often squirm around near humans, allowing bedbugs to leap onto us and spread their dangerous cargo!

The trouble with MRSA starts when it grubs around deep inside the body. MRSA is just too tanky to be killed by most antibiotics. This staph scoundrel is entirely resistant to a whole class of drugs called beta-lactams, which include penicillin, amoxicillin, and many more. Regardless of what drug is used, doctors must keep close tabs on MRSA patients. But don't start getting the shivers just yet! In March 2021, the FDA approved a new drug to treat MRSA: oritavancin. Oritavancin and a scanty band of other drugs like linezolid, daptomycin, and vancomycin are valiantly facing the MRSA onslaught. However, these drugs have their caveats. Patients treated with linezolid suffer from blisters, hives, and inflammation. Daptomycin triggers nausea, vomiting, and diarrhea.

Udderly Undesirable

To top it off, *S. aureus* infections aren't exclusive to humans. Milking cows suffer from *S. aureus* infections called mastitis, hindering their ability to produce milk. To limit the spread of *S. aureus*, farmers first milk healthy cows, then older cows and calves, and finally animals with an unknown infection status. Cows suffering from mastitis are given a break from milking. The dairy industry hates mastitis outbreaks—antibiotics used to treat infected cows can contaminate their milk. Anyone who drinks a glass would receive a nasty surprise! Poor Bessie.

Contracted by the Author

Staph aureus is an extremely nasty microbe that causes all sorts of grisly infections. It's so important to make sure our body is in tip-top shape to fight off disease. Working out, eating your veggies, or just simply taking some time to relax and blow off steam is essential to keeping your immune system sharp. Wounds should be treated immediately before *S. aureus* takes over the scene. Simple habits can have a huge difference.

Did you know yours truly also contracted a *S. aureus* infection? I caught the infection when I was around four, and it was *nasty*. A particularly adventurous mosquito stung me in the face, leaving a vicious itch near one of my eyes. When I scratched the itch, I cut myself over again and again and allowed *S. aureus* to march under my skin. Soon, the bump near my eye ballooned to approximately the size (and color) of a small tomato, squeezing my eye shut. I had to take an antibiotic called cephalexin to help my body bounce back. I also vaguely remember the doctor plastering a Band-Aid near my swollen eye! It also happened to be picture day at my preschool, so the Band-Aid spoiled the picture. *S. aureus* is an extremely pesky microbe...and is good at ruining school pictures as well!

CHAPTER 11

IT ALL STARTED WITH A RUSTY NAIL

IF YOU'VE GOTTEN A BIG CUT FROM A SLIP N' FALL, THE PAIN AND worry are usually soothed by a bit of antibiotic cream and a squeeze from Ma. But if you left the wound as is—no shots, no antibiotics, nothing—you might get a nasty *Clostridium* infection. *Clostridium* is a group of unique bacteria. When times are tough, *Clostridium* wraps itself up in a tough, outer shell called a spore. These spores are a force to be reckoned with. Heat and cold don't faze 'em. Antibiotics and alcohol? They can take it. Radioactivity? These little guys are up for the challenge! Spores have been *Clostridium's* secret recipe for success for millions of years.

But *Clostridium* has one Achilles heel: fresh air. Oxygen is kryptonite to these bacteria, so *Clostridium* snuggles down in damp, dark, airless places such as heaps of animal dung, globs of sewage, the digestive tract, and canned foods. (Baked beans, anyone?) Most organisms like me and you rely on oxygen to fuel our cells, but this is obviously not an option for *Clostridium*. Instead, these bacteria break down carbs in a process called fermentation, which does not require oxygen. In its stuffy, sultry homes, *Clostridium* multiplies normally and produces lethal toxins. Among the hundreds of species of *Clostridium* bacteria, the medical world is particularly worried about four of these little rogues: *C. tetani*, *C. botulinum*, *C. perfringens*, and *C. difficile*.

Sweet yet Deadly

All four of these microbial crooks don't just grow in animal plops; they thrive in honey and spoiled food. The reason why children younger than one can't slurp up raw honey is because their immune systems aren't strong enough to fight off the *C. botulinum* spores naturally found in honey. These rogue spores grow in their stomach and release botulinum toxin, *the deadliest toxin on the planet*, period. Fortunately, grown-ups can safely chow down on honey. *C. botulinum* enters the bodies of older, beefier people when they eat contaminated canned goods or when the bacteria enter a cut. Without treatment, *C. botulinum* overwhelms the body with its lethal toxin. Symptoms appear a few hours to a few days after infection and are mild at first (e.g., dry mouth, nausea, drooping eyelids, blurred vision) but soon progress to stomach cramps, vomiting, strained breathing, facial weakness, and paralysis. Yikes.

Some folks out there are even worried *C. botulinum* will be used by terrorists to inflict mass casualties. There are misassumptions that both Allied and Axis countries (unsuccessfully) used the bacteria as a bioterrorism agent in World War II. However, *C. botulinum* **can** serve as a biological weapon due to its extreme potency. But before we dig in further, I want to make this clear: no case of botulism in the US has been linked to bioterrorism. If mad scientists decide to shroud the world in lethal toxins, countries are prepared. Cases of botulinum toxin poisoning are hard to sniff out because symptoms of *C. botulinum* are similar to other diseases. But once the attack is discovered, health officials will spread the word and whip out a massive pile of botulism antitoxin (more on that later) from the Strategic National Stockpile (SNS) and dole it out to neighboring countries. People will be informed about how to identify botulism poisoning, and the Laboratory Response Network (LRN) will offer a wide array of botulism toxin tests to help people diagnose the disease. Medical professionals are even synthesizing potential botulism antitoxins right now! In the unlikely event of a bioterrorism attack, we're ready.

Despite *C. botulinum's* destructive potential, this bacterium is used in the cosmetics industry. Botulinum toxin—or Botox—blocks the nerve signals controlling muscle activities. This normally leads to paralysis, but when given in minute amounts, it relaxes muscles. Botox removes wrinkles, eases bladder trouble, and quells headaches, but it also helps mitigate more serious illnesses like cerebral palsy (a neurological disorder that causes excessive muscle contractions) and lazy eye (where the muscles of the eye don't work properly). Botox even lifts people's moods! Ever feel happier after you raise your eyebrows, or angrier when you scowl? Scientists believe Botox reduces anxiety by freezing facial muscles, causing people to feel at ease. In fact, studies have shown that Botox injections can reduce anxiety by as much as seventy-two percent!

Watch out for Rusty Nails

If you've ever heard rusty nails spread a deadly disease called tetanus—caused by *Clostridium tetani*—this is correct. However, the nail does the damage, not the rust or dirt. Whenever you accidentally stab yourself with a nail or any other sharp object, you open the perfect pathway for *C. tetani* spores to enter the body. The irony here is that *C. tetani* are shaped like nails! In the presence of low-oxygen conditions, the bacteria multiply, sop up nutrients, and release their toxins. The toxins are dispersed through the bloodstream and affect the body's neurotransmitters. Neurotransmitters are chemical bridges that relay signals between nerves. Any interference here is *bad*. This first results in sudden muscle contractions, and later, paralysis. One of the first symptoms of a *C. tetani* infection is lockjaw. Lockjaw is when the jaw muscles seize up and lock in place. This is dangerous enough, but if left untreated, paralysis caused by *C. tetani* gradually descends down the body, eventually suffocating a patient. This is known as generalized tetanus. Without the protection of a vaccine, the chances of dying from generalized tetanus are about thirty-five percent!

A less severe but less common form of tetanus is localized tetanus. This infection occurs when tetanus bacteria linger in the infected area and causes persistent muscle contractions. Localized tetanus sticks around for weeks and can be a warning that a more severe tetanus infection is on the horizon. A rare infection called cephalic tetanus arises when *C. tetani* invades the ear and affects the facial nerves after a blow to the head.

Holiday High Jinks

C. perfringens and *C. difficile* produce toxins that trigger bouts of stomach pain and diarrhea. These bacteria love growing on turkey and roast beef. *C. perfringens* outbreaks frequently occur in November and December because of the traditions of munching on holiday turkey. Most *Clostridium* conquests happen on food put in the "danger zone." The danger zone is a broad range of temperatures inadequate for storing food, from a wintery forty degrees Fahrenheit to a blistering 140 degrees Fahrenheit. At these temperatures, pathogens can take over and spoil your dinner if left outside the fridge for more than one hour. I guess bacteria celebrate Christmas too! Most cases of *C. difficile* and *C. perfringens* are mild, no worse than a typical stomach bug. The human microbiome naturally crowds these naughty fellas out.

However, some of the cases of this bacteria are, um, more dramatic. *C. perfringens* causes a lethal condition called gas gangrene that occurs on wounds and inside the intestine. Gas gangrene is very rare and mostly affects people with weakened immune systems. During gas gangrene, *C. perfringens's* toxins kill the body's soft tissue, creating a noxious gas that emits a horrendous odor. Gas bubbles trapped within the body even make ominous crackling noises. Gas gangrene can affect the body within *minutes*.

Without treatment, death is certain. Even with the help of antibiotics, *C. perfringens* can barricade itself in dead tissue, quickly rendering treatment ineffective.

C. difficile stirs up the tongue-tying pseudomembranous colitis. It is a potentially deadly complication that typically affects the elderly. Pseudomembranous colitis (phew, what a word!) provokes severe intestinal inflammation and even causes the intestine to leak pus, potentially resulting in dehydration. This disease can be cured, but reoccurs as often as twenty percent of the time. However, pseudomembranous colitis is only the tip of the iceberg. Some people suffering from pseudomembranous colitis develop toxic megacolon. Toxic megacolon takes place when the intestines are overwhelmed with toxins and balloon to massive sizes. These inflamed intestines can even rupture, leading to a slow, agonizing death. Thankfully, this deadly condition is very rare.

Back in My Day...

Clostridium food poisoning was a big problem back in the olden days. Good thing cooking destroys *Clostridium* toxins. However, *Clostridium* spores were a tough nut to crack. The ancients had no way to safely preserve food; the best they could do was to douse their nosh in salt, freeze their grub in the snow, or dry it in the sun. In ancient times, people often outlawed selling or eating meat prone to being contaminated with *Clostridium*. Leo VI, emperor of the mighty Byzantine empire, banned his people from eating blood sausages, fearing they would contract *C. botulinum.*

C. botulinum didn't get its present name until 1870, when German physician John Muller coined the disease "botulism" from the Latin word, *"botulus,"* or sausage. (Coincidentally, *C. botulinum* also looks like a sausage.) A botulism antitoxin was discovered a century later during the 1970s after researchers extracted botulinum toxin antibodies from a horse named First Flight. Within twenty years, more antitoxins were developed. Antitoxins drastically improved the outlook of botulism patients. Previously, half of all infected victims died from botulism. But when treated with antitoxin, more than 99.5 percent of infected people survive. However, antitoxins don't reverse botulism paralysis; they only stop the damage. It can take weeks, months, or even years for people to recover. Worse, the paralysis even opens opportunities for secondary bacterial infections to siege the body.

Tetanus has a bit of a different story to tell. Records show it has poisoned people since ancient Romans paraded around in togas. Fortunately, tetanus is a bit less deadly than botulism, killing about fifteen to twenty percent of infected victims. In 1884, the bacteria were proven to cause an illness after pus containing *C. tetani* was injected into animals (gross!). Tetanus scourged humanity until 1889, when Kitasato Shibasaburo—a Japanese bacteriologist—isolated the bacteria and neutralized its toxin with antibodies. In 1897, Edward Nocard and Emil von Behrig discovered tetanus antibodies passed down from person-to-person KO'd *C. tetani*. During the 1920s, researchers realized formaldehyde inactivated tetanus toxin but provided immunity against it—eventually leading to the tetanus vaccine. But scientists didn't stop there! In 1926, Alexander Glenny and his troupe of scientists found out a chemical salt called alum beefed up the tetanus vaccine. And in the 1940s, people successfully mixed diphtheria and whooping cough vaccines with tetanus shots—a three-in-one package! These vaccines saved countless lives and allowed us humans to reach ripe old ages.

The tetanus vaccine utilizes inactivated tetanus toxins to spark an immune response. Your loyal troop of white blood cells then unleash antibodies to neutralize the toxin. However, this immunity doesn't last forever. You probably received the diphtheria, tetanus, and pertussis (whooping cough) vaccine (DTaP) when you were a baby. Your final dose of DTaP was presumably given to you when you were a wee lad of four to six years. A different vaccine called TDaP is given to boost your immunity against these three diseases when you are eleven or twelve. When you start paying taxes and chugging coffee, you will receive a dose of TDaP every ten years to keep your immunity fresh and strong.

Adding Fuel to the Fire

Unlike *C. tetani* and *C. botulinum*, doctors get headaches when treating *C. difficile*. When patients gulp down antibiotics, the antibiotics destroy their microbiome. This allows *C. difficile* to invade the gut. Normally, most people trounce *C. difficile*. But for patients relying on antibiotics to treat a chronic infection, *C. difficile* infections are a sticky situation. Ironically, this calls for more antibiotics! Two drugs—metronidazole and vancomycin—stuck around to clobber *C. difficile*...until the early 2000s. At the turn of the millennia, a hypervirulent *C. difficile* strain called NAP1/027 entered the arena. NAP1/027 is resistant to **all** antibiotics and secretes massive amounts of toxins, even by *Clostridium* standards. New antibiotics like fidaxomicin were made, but NAP1/027 eats them for breakfast!

Antibiotics are clearly not an option when treating *C. difficile*. To cross swords with this pathogen, scientists genetically modified a strain of *C. difficile*, disabled its toxin, then implanted the weaker microbes back into patients' guts. These implants of harmless *C. difficile* create a "family feud" between the strains of *Clostridium*, reducing the amount of nutrients and space available for the infectious strain. After a *C. difficile* infection is treated, the bacteria often bounce back and take over the intestines once more. But when pressed against the GMO'd *Clostridium* strain, the recurrence rate of infectious *C. difficile* dropped from thirty-three percent to eleven percent. Eating probiotic-rich foods like yogurt and cheese also helps crowd out *C. difficile*.

However, the most effective way to bash heads with *C. difficile* is to replenish the body's gut flora, or good microbes. A Dutch doctor beefed up his patients' microbiomes by injecting watered-down feces into their intestines. Disgusting...but when treated with feces from healthy people, eighty-one percent of patients were on the fast track to recovery. Antibiotics, on the other hand, only cured thirty-three percent of infected patients. Injecting someone else's, ahem, waste into your digestive tract seems gross, but fecal transplants create no adverse side effects and are much better (and cheaper) alternatives to antibiotics. Just don't try this at home!

Clean, Separate, Cook n' Chill

Clostridium infections are miserable, so it's best to scooch away from 'em. Vaccines always help, so make sure your shots are up to date. If you receive a deep wound, it's best you get another tetanus shot if you haven't recently had one. To shy away from botulism, don't let babies younger than one snack on honey! Even pasteurized honey is dangerous for them. You just can't kill *C. botulinum* spores! In fact, four babies who sucked on honey pacifiers were almost killed by botulism. According to the CDC (Centers for Disease Control and Prevention), "Honey may quiet them, but botulism may kill them!" Botulism is also linked to illegal drug injections. Many people who contract botulism take illegal drugs. Contaminated drug syringes and needles pave a direct route for *C. botulinum* to the body. Sharing needles also allows *C. botulinum* to jump from place to place. *C. botulinum* symptoms are like opioid overdoses, throwing medical professionals for a loop. Please, don't take illegal drugs!

To prevent stomach trouble from *C. perfringens* and *C. difficile*, clean, separate, cook, and chill your food. "Clean" means to give your grub (and whatever is used to prepare it) a good rinse. "Separate" means keeping equipment used for preparing/storing uncooked meat, poultry, eggs, and seafood *far* away from utensils used to conjure up your Caesar salad. "Cook" is self-explanatory; grill meat at temperatures sufficient to kill all pathogens. Bear in mind seafood needs to be cooked at 145 degrees Fahrenheit, red meat at 160 degrees, and poultry/leftovers at 165 degrees. "Chill" is vital too. Cooked food should be stored in a fridge below forty degrees Fahrenheit (my fridge stores food at thirty-seven degrees Fahrenheit) and in the freezer at or below zero Fahrenheit. To speed up the chillin', evenly divide food into chunks. And when your leftovers start to get a bit "off," throw them out!

It's Everywhere!

All in all, *Clostridium* is a ubiquitous yet dangerous pathogen that has wreaked havoc on society since as long as we can remember. It is important we remember *Clostridium* is lurking everywhere. So get those shots, prep your food properly, keep your gut happy by eating healthy foods...and watch out for those rusty nails!

CHAPTER 12

THE BIG BAD POX

SMALLPOX. THE DEADLIEST VIRUS ON THE PLANET HAS KILLED half a billion people just in the twentieth century alone. It has toppled mighty rulers, vanquished courageous generals, and trounced valiant chieftains. Whenever empires expanded, explorers sailed across oceans, or when merchants marched across deserts, smallpox followed like a malicious shadow. And yet, the virus is gone. Kaput. Defeated. Squashed, if you will. The last samples of smallpox are awaiting their destruction in high-security labs.

All this attention smallpox has gotten over the last couple of centuries is probably making its cousins—other poxviruses like camelpox and monkeypox—jealous. After all, camelpox causes lymph nodes to balloon and blacken all over a person's body. Monkeypox (renamed mpox since monkeys are not the source of the disease) is capable of causing big outbreaks. It frazzles humans with painful rashes and high fevers. Nevertheless, smallpox is still the king of all poxviruses, killing thirty percent of all unvaccinated victims.

Despite its lengthy reign of terror, smallpox is a relatively new virus. Back in the day when humans pushed mammoths off cliffs, smallpox was nonexistent. However, one hot evening in Africa 68,000-15,000 years ago (perhaps on a Saturday night), the great-great-great-great grandpa of smallpox infected a desert rodent. This squirming little furball infected another rodent, and a milder version of smallpox spread among these mousy mammals. A few centuries later, this proto-smallpox evolved into three other viruses: camelpox, which still clobbers camels today; taterapox, widespread among rodents in the Middle East; and the ancestor of modern smallpox, which infected humans.

Smallpox could've also emerged from Europe. A mouse could've infected a *very* unlucky human with a smallpox-like virus thousands of years ago. Before it became the smallpox we know today, this virus split into two species: one kept becoming more fatal and virulent until it turned into smallpox. The other was much milder and less transmissible, eventually becoming extinct in the brutal hands of smallpox.

What's in a Name?

With all the juicy history out of the way, let's talk about some trivia. First, you may be wondering why such a deadly disease is called "small." Doesn't a more pompous name like "Great Pox" or "Killer Pox" sound better? Well, there was already a Great Pox around—a completely unrelated infectious bacteria called *Treponema pallidum*, or syphilis. To clearly distinguish the virus from syphilis, people dubbed the disease "smallpox." Second, there are actually two variants of smallpox: variola major and variola minor. Variola major is the head honcho, the big deal. It kills about thirty percent of its infected victims. Variola minor is more like variola major's wannabe; trailing behind its *way* more prominent cousin, it only kills one percent of the people it infects.

A Globetrotting Virus

As early as 200 BCE, smallpox walloped Egyptian pharaoh Ramses V. In 300 CE, the ancient Chinese pleaded to the god of smallpox, Yo Hoa Long, to cure the terrible disease. By the sixth century, increased trade with China and Korea allowed smallpox to leap onto the shores of Japan. Medieval Japanese wore bright red to "scare away" smallpox and built shrines to satisfy the "smallpox demon's" bloodlust. Just a mere 100 years later, smallpox marched into India. Indian smallpox goddess Shitala Matla was believed to both cause and cure smallpox and motivated people to keep up their hygiene. Not easy for a society in ye old ages! Simultaneously, as the Arab empire began to expand, smallpox trooped into Spain, Portugal, and Northern Africa. In the tenth century, smallpox hitched a ride on the silk road to Turkey. In the eleventh century, the Crusades blasted arrows and cannonballs (as well as the smallpox) further into Europe. By the thirteenth century, smallpox even wintered it out in Iceland. And in the 1400s, Portugal claimed part of West Africa. Predictably, smallpox conquered West Africa along with the

Portuguese. When the 1500s rolled around the corner, the slave trade introduced smallpox to the steamy tropics of South America and the Caribbean—another reason why slavery is so horrible and atrocious. When the seventeenth century kicked in, European explorers plonked smallpox down in North America and Australia.

Smallpox tore across the globe like wildfire!

Just Plain Horrible

Smallpox is a harrowing disease. It stays hidden in the body for seven to nineteen days before assailing its victims with agonizing headaches, high fevers, and bouts of vomiting. Red spots explode all over the mouth and tongue and sores break out over the skin. Paradoxically, smallpox patients often feel *better* as the rash develops...then their health rapidly declines as their skin sores fill with pus. Death usually occurs by the fifth day of the rash. Those lucky enough to dodge the grim reaper are covered with pustules (hard, round, lesions that feel like peas under the skin) for about ten days. The pustules then scab up and after about six days, they fall off, leaving patients covered in scars and pockmarks due to the destruction of their sweat glands. In the worst cases, smallpox survivors suffer brain inflammation, bone infections, and even blindness.

Unfortunately, this is the most *common* type of smallpox. Like some sort of nightmarish Pokémon, smallpox has two more forms that are particularly concerning. Flat-type (or malignant) smallpox causes lesions to merge and form a flat, soft rash. This "fuzzy" rash is an extremely bad sign a person's immune system is not tussling with the virus, allowing the virus to take over the body unopposed. Flat-type smallpox is more common in children but is extremely lethal. Worst of all is hemorrhagic smallpox. Though it only accounts for three percent of all smallpox cases, hemorrhagic smallpox is guaranteed death. The skin of hemorrhagic smallpox patients bleeds, giving them a burnt, charcoal-like appearance. Hemorrhagic smallpox also triggers massive internal bleeding and organ failure; victims often die from overwhelming amounts of pathogens in their blood.

Jenner Saves the Day

As smallpox caused rashes, deaths, and more rashes, a tiny spark of hope emerged from the ashes. In the 1700s, a young, aspiring scientist named Edward Jenner noticed that milkmaids infected with cowpox were immune to smallpox. To test his theory, he took a cowpox sore from his milkmaid, Sarah Nelmes, and inoculated the sore into the arm of James Phillips, the son of Jenner's gardener. Inoculation is a fancy, more concise, and more polite way of saying, "to inject infected material into someone's body." Jenner then injected smallpox into James's bloodstream, but James came out unscathed. He had invented the smallpox vaccine! However, this fame came with a fair share of controversy. People didn't like that some mad scientist was injecting cow sores into their blood.

Despite some public opposition, smallpox vaccines became widely used and saved millions of lives. Later, the vaccinia virus—a harmless cousin of smallpox—was used as a vaccine instead. But even before Edward Jenner, inoculation had already become a common practice in many areas around the globe in the 1000s. Inoculation caused high fevers, swollen armpits, and insomnia, but was a far cry from the lethal onslaughts of smallpox. Rarely, the virus vaccine induced painful allergic reactions and even killed immunocompromised patients, one of the caveats of using a live virus as a vaccine.

Now It's Gone

The WHO (World Health Organization) devised a plan to eradicate smallpox in 1959. Though smallpox disappeared from North America in 1952 and in Europe in 1953, the plan never caught on, on a global scale. Smallpox continued to mercilessly thrash humanity until 1966. But in 1967, the eradication plan was intensified, and the WHO launched into action. Laboratories churned out high-quality smallpox vaccines. Scientists spruced up the smallpox vaccine, so it was as effective as possible. Soon, smallpox was kicked out of South America in 1971, Asia in 1975, and Africa in 1977. But things are never so simple. In 1978, smallpox escaped from the confines of a lab and killed medical photographer Janet Parker. Lab workers and the public alike were horrified, and 500 people were vaccinated in a desperate scramble to prevent the virus from trickling back into the world once more. Smallpox wasn't declared eradicated until the 1980s.

Why was smallpox, one of the deadliest diseases of all time, eradicated? Though this virus seems like a formidable final boss, this pathogen has received numerous nerfs. An effective vaccine was available to conk out smallpox. It has no animals it can fall back on and infect once its human hosts develop a vaccine. And perhaps most importantly, smallpox was only contagious after people developed symptoms, allowing doctors to quarantine and treat patients.

Walk of Fame

Despite smallpox's appalling and deadly infections, many people survived contracting the virus and lived to tell the tale. Presidents George Washington and Abraham Lincoln duked it out with the disease...and won. (George Washington recovered in just a month.) Piano prodigies Beethoven and Mozart conquered smallpox, too. Mary Shelley, the author of *Frankenstein,* survived as well. And don't forget Queen Elizabeth I, who clobbered smallpox at age twenty-nine.

However, these survivors are lucky. Countless people and even entire civilizations were wiped out by smallpox. Two Chinese emperors of the Qing Dynasty (1636-1912) named Shunzhi and Tongzhi both died because of this dreadful disease. They were not alone. The mighty Incan and Aztec empires perished from small-pox after Spanish conquistadors struck their kingdoms with the mortal blow. Pocahontas (possibly) died from the disease shortly after she sailed to Europe. Entire tribes of Native Americans died from smallpox after European colonists brought the disease to the Americas. One of the many reasons why Indigenous People's Day replaced Columbus Day is because Columbus and his crew introduced smallpox to America.

There is some evidence smallpox was even used as a bio-weapon. Colonists learned indigenous peoples were extremely vulnerable to smallpox due to their lack of immunity toward the disease. Some settlers even said smallpox was "more deadly than an enemy weapon." Colonists warring with Native Americans gave them "gifts" of blankets infested with smallpox in the hopes of killing them!

Close Calls

Despite all the carnage it has caused, smallpox is a thing of the past. However, people are still debating whether to destroy its last samples. Some think it's best that test tubes full of the deadly virus be kept intact for future research. Others want to destroy the virus forever. People fear rogue countries and terrorist groups will use smallpox as a secret weapon, or a lab might accidentally release smallpox back to the planet. The risk of a lab leak is a very real danger and could have occurred many times.

In 2013, cloned smallpox fragments were found in a South African laboratory. The WHO immediately killed the viruses. In an FDA lab, employees unearthed vials containing smallpox. These vials were quickly transferred to the CDC and locked up, awaiting their destruction to this day. In 2019, a Russian lab holding small-pox samples experienced a gas explosion. This injured one worker but fortunately occurred far away from the area containing small-pox. In 2021, several vials labeled "smallpox" were discovered in the freezer of a vaccine company. People soon found out these vials only held the vaccinia virus.

What do We do Now?

At the end of the day, smallpox isn't just the world's deadliest virus. It has participated in wars, wiped out civilizations, and changed history. It kickstarted the most devoted vaccination project ever. Now, the reign of smallpox only stretches into the icy recesses of laboratory freezers. But what should we do next? Should we blast the last smallpox samples with a laser like the Death Star did to Alderan? Should we keep the samples for further research? Or do we just dig a deep hole in a desert, lower the vials in, and cover them up? Some people think that smallpox is being developed as a terrorist weapon. A lot of messy politics come up, which is too overboard for this humble book to explain. But really, we have just won a battle in a series of wars against microbes. In only a few centuries, there could be a new smallpox-like virus terrorizing the world again.

But we can fight back with our vaccines and prior knowledge. There are also new drugs developed to treat the unlikely event of a smallpox infection, such as the antivirals tecovirimat and cido-fovir, not to mention our old buddy the smallpox vaccine, which hasn't retired yet. Our old friend fought the mpox outbreaks that occurred from 2022-2023. Still, new diseases keep blossoming from the ashes of old. Nature keeps building, keeps creating. It's neither good nor evil; it's just another reality we must accept on our wonderful planet.

CHAPTER 13

FISHY THINGS AND SEAGULL WINGS

FISH ARE MARVELOUS CREATURES. THEY RANGE FROM THE TINY sign eviota, which lives for a paltry eight weeks, to the Greenland shark, which can celebrate its 500th birthday beneath the frigid Arctic Ocean. They are blisteringly fast like the sailfish, which speeds through the water at highway speeds, or lethargically slow, like the dwarf seahorse, which reaches a measly 0.001 mile per hour at a full sprint. However, all these spectacular creatures are susceptible to a disease called *Mycobacterium marinum*.

Bacterial Buccaneers

M. marinum is a salty seadog of a bacteria that usually affects fish and amphibians but can infect people too. These infections result from bacteria sneaking into a cut or scrape but are also prompted by exposure to infected animals.

M. marinum isn't just found in fish tanks or the blue briny: the bacteria also flourish in dirty swimming pools. However, *M. marinum* doesn't have to stay in the water! While lounging on a beach in Costa Rica, a three-year-old girl contracted the bacteria when she was bitten by a greedy iguana trying to nab a piece of cake in her hand. She developed a small, red bump (known as a granuloma) on her skin. Granulomas form when white blood cells mob and dogpile bacteria, creating a dark red mass of infected tissue. Other infected patients develop itchy rashes and sores. *M. marinum* infections are usually not lethal but are quite irritating. But in immunocompromised patients, *M. marinum* infections are a different story. These unfortunate souls suffer from fevers, chills, swollen lymph nodes, lung infections, bone infections, and even arthritis. These infections are even potentially lethal to them. The long, slow incubation period of *M. marinum* makes it tricky to diagnose, and this scurvy knave of the seas is often confused with other diseases.

But fish and other aquatic life have much more to fear from *M. marinum* than us landlubbers; this microbial swashbuckler is even called "fish tuberculosis." Remember the seahorses I mentioned earlier? They are extraordinarily and even notoriously susceptible to *M. marinum*. The bacteria are a frustrating nuisance to people who keep these exotic fish in their aquarium. Sea bass (delicious!) are also commonly affected by this disease. Many freshwater aquarium fish are also prone to *M. marinum*. The bacteria rip gruesome lesions all across their bodies and can kill up to 100 percent of some infected fish. When the unlucky fish kicks the bucket, *M. marinum* is released from the fish's internal organs.

When other fish scavenge or swim near the contaminated carcass, they become infected as well. *M. marinum* was a huge nuisance to salmon farming. Salmon fed contaminated scraps of dead fish (or offal) died in massive droves. Because of economic damages and ethical problems, using offal as fish food has been discontinued. Dolphins, belugas, manatees, and seals kept in suboptimal conditions are also at risk for contracting *M. marinum*, another reason why captive marine mammals are not a good idea!

Fortunately, *M. marinum* infections in humans are extremely rare. The chances of getting infected from the disease are a mere one in 3,700. People most at risk of contracting *M. marinum* are those continuously exposed to contaminated water, such as fishermen, aquarium workers, and fish keepers. Though obnoxious and itchy, this micro-buccaneer isn't contagious from person to person. The risk of catching *M. marinum* doesn't mean you have to flush your fish down the toilet, but it's a good idea to check if your fish tank is squeaky clean. When handling a fish, use gloves and wash your hands afterward. If you like a nice tilapia filet and are spooked by munching on fish, don't worry! *M. marinum* is killed at high temperatures or when dunked in sushi vinegar.

Flatworm Freebooters

Another "fishy" disease is clonorchiasis, a parasitic infection caused by the flatworm *Clonorchis*, aka the oriental liver fluke. There are many types of liver fluke: some lurk in the depths of the ocean while others patrol the recesses of lakes and rivers. All are actually tiny animals called flatworms that thrive in the muscles of fishes. Vegetarians aren't safe either: some liver flukes squelch around on freshwater plants. *Clonorchis* is commonly found in freshwater fish in East Asia. Raw or undercooked fish are often the culprits of *Clonorchis* infections. However, *Clonorchis* is one stealthy parasite: most people who are infected don't show symptoms...at first. Infections often persist over the twenty-five- to thirty-year lifespan of the parasite and are potentially fatal, but some lucky people get over this flippant flatworm just fine.

But when *Clonorchis* strikes, things are not good. The parasite burrows into the liver and the gallbladder. The liver helps your body filter toxins and absorb nutrients while your gallbladder secretes bile—a greenish liquid that helps break down and absorb fats from food. When these organs are going down, things turn gnarly. *Clonorchis* evokes acute hepatitis (severe liver inflammation), gallbladder obstruction, diarrhea, abdominal pain, fatigue, malnutrition, and nausea. Sometimes, people's skin will turn a sickly hue of yellow; a sign they are suffering from jaundice.

Lightly cooked, pickled, or smoked fish can still house this parasite, so cook your salmon steaks properly. When *Clonorchis* wriggles into your body, a proper dose of nifty antiparasitics knocks it out easily. To diagnose this freshwater swashbuckler, a feces sample sent to a lab will always do (Yuck!). Scientists use an ultrasound, CT, or MRI to detect *Clonorchis* eggs. These eggs are distinguished from other parasites by examining the microscopic features on them. However, this method has its limits: when the parasite's lifespan is approaching an end, feces examination is less and less effective. A less repulsive method to scan for *Clonorchis* is to test for this flatworm's antibodies in the bloodstream. But this testing is wonky, ineffective, and unavailable in the US.

Speaking of poop, feces are a very important part of the *Clonorchis* life cycle. When an organism infected by the parasite excretes, *Clonorchis* eggs are released into the environment. Then, an important intermediate host comes along: freshwater snails. These snails accidentally ingest the eggs and give the parasite a safe place to grow. First, the eggs hatch into baby parasites with tiny swimming hairs on the sides of their body. These flower-like organisms are called *miracida*. The *miracida* morph into small, blobby *sporocysts*, which transform into avocado-shaped *rediea*. Despite their plump appearance, *rediea* can clumsily swim through the water. These rotund orbs change into *cercariae*, which look like tadpoles with beards. The *cercariae*, ahem, exit the snail and loiter in the water looking for their next hosts: fish. After latching onto a fish, the *cercariae* contort into *metacercariae*, which curl up inside the fish's gut or skin. When humans or other seafood-loving animals gobble down the infected fish, the *metacercariae* develop into adult flukes and lay their eggs. Cattle and deer are also potential victims of *Clonorchis* if they muck around in wet, marshy areas. After the eggs exit the body through feces, the cycle continues.

Sea Star Slime Plagues

Another marauder of the high seas is sea star wasting syndrome. This disease doesn't affect humans, fishes, or whales, but wipes out entire *populations* of sea stars. The first signs of sea star wasting syndrome in sea stars are the appearance of white lesions on their skin. Infected starfish also lose their appetite. The limbs—then the rest of their bodies—of suffering sea stars slowly decays into a white, gruesome-looking goo. These gigantic "sea star slime" plagues have been dated back since 1978. Sea star wasting syndrome always kills more sea stars dwelling in warm water than in cold water. The exact cause of the disease isn't clear, but there are a few hypotheses. A type of nutrient-loving (or copiotrophic) bacteria thrives in warm water and lives in the same habitat as sea stars. The bacteria grow and multiply on the starfish and gobble up oxygen from the water, suffocating the sea stars and causing their tissues to die, hence the white goo. These hangry copiotrophs aren't infecting the sea stars but *are* starving them from nutrients and oxygen. Other possible culprits for sea star wasting syndrome are viral infections, haywire microbiomes, and fungi. Environmental factors such as freshwater runoff, oxygen depletion, etc. may also cause sea stars to bite the bullet. Global warming is often traced to the root of the problem as warming oceans promote copiotroph overgrowth. Worse, as more sea stars die, more nutrients are available to the copiotrophs, creating a deadly cycle of exponential growth.

To add to the sea star wasting syndrome puzzle, not all starfish succumb so easily to this disease. Blood stars and leather stars weather the disease as neighboring starfish shrivel up into puddles of decaying tissue. Sea stars with flashy spikes and protrusions seem to be more susceptible to the disease than plain sea stars because more bacteria can grow on these more heavily ornamented starfish. This disease is one of many damaging effects of global warming finally being brought to light!

Loooooooong Tapeworms

Once upon a time a man pulled a five-and-a-half-foot-long tapeworm out of his body. This story is no joke, but seems to defy science. Gigantic tapeworms are definitely not microbes, but we'll give them a special mention here since they start out as tiny eggs. After suffering from bouts of dysentery—microbiology speak for "bloody diarrhea"—the man decided he had enough. On one fateful day, he walked into a hospital and requested to be treated for parasitic worms, or helminth. Doctors thought the man was joking until he pulled out a wad of toilet paper with a tapeworm wrapped inside. He had pulled it out during an "extremely disturbing defecation episode". At first, the man thought his intestines were flowing out of his body. Then, he saw that monstrosity of a tapeworm. After all he had been through, it's no surprise the man was relieved his organs weren't falling into the toilet. How did the tapeworm squiggle its way into his body?

Turns out the man had a sushi obsession, especially with salmon sashimi. Despite the meticulous preparation of sushi, things can go wrong. His sashimi, prepared from Alaskan-caught salmon, contained a fish tapeworm named *Diphyllobothrium latum* usually found squirming around Japan. Once ingested, the tapeworm scooted its way into his gut and caused havoc. And while six feet long is astoundingly large for a tapeworm, some grow to thirty or forty feet. And in whales, tapeworms can grow up to **120 feet.** (Gross!) Tapeworms are not the only parasitic worm that can reach uncomfortably large sizes: another type of enormous, water-borne helminth is the guinea roundworm, which commonly affects impoverished villages. Transmitted by freshwater micro animals called water fleas, guinea roundworms grow up to three feet in length and emerge through a blister in the skin. The worm often has to be reeled out for several days with a stick. Imagine having such a large monstrosity wriggling inside you!

However, *D. latum* infections are very rare. Most cases are in Asia, but migrating fish sometimes ferry the worm to South America and Alaska. Marine fish such as herring and sea bass are most likely to carry this tapeworm. Farmed fish and mahi mahi (aka dolphinfish) are the least likely to have helminth squiggling around inside them. Unfortunately, *D. latum* is spreading fast across the wide ocean blue. But with fewer than 100 cases in humans per year, you can still indulge yourself with sushi...even if it is salmon sashimi.

More Worms?!

Fish tapeworms aren't the only sushi outlaws. A worm called *Anisakis* also hitches a ride on raw fish. *Anisakis* sticks to salmon, swordfish, squid, octopus, shrimp, and/or crabs. The *Anisakis* worm is transferred from fish to fish up through the food chain and eventually ends up in the big ol' guts of a whale. *Anisakis* is only found in the ocean, so when it worms its way into a human (pun unintended), it isn't adapted to living in such a host and wriggles itself to death after a few weeks. But before it kicks the bucket, this pesky worm burrows into the esophagus, stomach, and intestines, and it causes a huge bellyache. Many people suffer from nausea, vomiting, and even cough up *Anisakis* larvae. This provides a gross but convenient way of diagnosing *Anisakis*. Another way of testing is to have an endoscope (a camera specifically designed to examine the intestines) slide around your innards to search for worms. Yuck! Ironically, examining feces isn't helpful for diagnosing *Anisakis* because the parasite can't complete its natural life cycle nor produce any eggs in the human intestine. It just migrates under the skin and into the digestive tract...still not a big consolation to anyone suffering from this parasite!

Raiders of the Blue Briny

All in all, "land diseases" get all the attention. But pathogens that are found in lakes, oceans, and rivers are just as deadly as their landlubber counterparts. It is important not to underestimate the power of aquatic diseases, especially those that can knock out thousands of sea stars in just a few weeks. These diseases deserve a little more attention. After all, no one wants a forty-foot-long tapeworm wriggling inside their intestines!

CHAPTER 14

TB OR NOT TB?

When saber-toothed tigers prowled the Arctic tundra and mammoths thundered across grasslands 9,000 years ago, something strange was going on in the city of Atlit Yam, a prehistoric city now sunk underneath the coast of Israel. A mother and child fell dangerously ill. They suffered from chest pains, chills, fatigue, and coughed up blood. Spongy yellow lesions called tubercles erupted through their lungs, which the mother and child coughed up. Eventually, they died and received a tearful burial. The lethal disease that claimed their lives? Tuberculosis, or TB for short.

This Old-timer Means Business

The earliest case of TB is traced back to Atlit Yam but is unlikely to be the origin of the disease in humans. Tuberculosis has hassled life on Earth for a very, very long time. The ancestors of tuberculosis-like bacteria originated 245 million years ago, infecting seafaring reptiles and, possibly, our rat-like mammal ancestors. Seventeen-thousand-year-old bison fossils show these prehistoric cattle suffered from "cow TB," or *Mycobacterium bovis*. Primates, deer, rodents, swine, cows, cats, dogs, seals, sea lions, and even guinea pigs (squeak) are all plagued by these devilish bacteria. The earliest written mentions of TB were jotted down 3,300 years ago in India and 2,300 years ago in China. And from the Middle Ages to the time George Washington commanded the Continental Army, TB was responsible for twenty-five percent of all human deaths.

TB cases skyrocketed in America and Europe during the Industrial Revolution. While jam-packed cities increased tech advancements, cramped towns spread TB like wildfire. Poor workers fared the worst: they lived in hideous and overcrowded conditions and were often malnourished. These weak workers' lungs were like theme parks for TB. At first, TB infections seem trifling. When TB first pours into a person's body, the bacteria are destroyed and vanquished by trusty white blood cells. This is called the primary infection; during a primary infection, most people feel completely fine or suffer from flu-like symptoms.

However, not all bacteria are trounced, and even when business in the body returns to normal, TB stealthily multiplies. If an infected person's immune system cannot control the bacteria, TB explodes across the body. This is known as active TB. The pathogens overwhelm the lungs. Mucus-producing cells called goblet cells try to help by clogging the lungs with sputum (the gunk that stuffs up your nose when you get a cold). This snot carries a brigade of valiant white blood cells that duke it out with the bacteria.

Unfortunately, the bacteria can seize the upper hand: infected patients lose weight and are besieged by nasty fevers and fits of coughing. Some unfortunate souls even collapse from fatigue in the middle of the day. In severe cases, patients develop swollen lymph nodes, which cause firm, reddish-purple lumps to fester. Like a (killer) fish out of water, TB can even spread *outside* the lungs and infect other internal organs, including the brain! Eventually, with a deadly combo of starvation and stress, many people during the Industrial Revolution died from TB.

TB "cures" given at the time were at best ineffective. Most doctors could only prescribe warmth, rest, and good food—*"Lana, letto, and latte,"* as said in Latin—to keep patients comfy. Some treatments were harmless, like a "purifying" touch from a queen or king. Some were a bit disgusting: a few doctors attempted to treat patients with smelly cod liver oil and vinegar massages. Others were downright dangerous: people were so desperate to treat TB that they inhaled poisonous plants like hemlock or smelled whiffs of turpentine, a gooey black substance that smells like decaying carcasses and has the potential to be an addictive drug.

Finding a Cure

Many renowned people died from tuberculosis without a cure. Famous naturalist David Thoreau, First Lady Eleanor Roosevelt, French inventor and educator Louis Braille, US president Andrew Jackson, and activist Nelson Mandela succumbed to TB. They aren't alone: in just the past two hundred years, TB has killed a billion people. Even today, TB claims 1.3 million lives each year. TB continued to ravage the world until brilliant scientist Selman Walkman discovered an antibiotic called streptomycin that could take on TB. Streptomycin saved its first life in November 1949 and punched out TB infections worldwide. Many more antibiotics were discovered to treat TB, and the survival rate of TB patients skyrocketed, especially when concoctions of multiple medications were used all at once. A vaccine for TB—invented in 1921 by Camille Guérin and Albert Calmette—known as BCG helped prevent infants and toddlers from contracting TB meningitis. However, the vaccine may not protect against TB lung disease. To make matters worse, immunity only lasts fifteen years, meaning people have to keep themselves up to date on vaccinations. Think about the last time you got a measles shot. Unless you are reading this as a five- or six-year-old, it probably was a pretty long time ago. Despite being crummy, BCG is still saving lives in countries where TB is rampant, such as India, China, and Bangladesh.

The Bacterium Strikes Back

Lives were being saved all around the world. In America, TB almost became a thing of the past. However, developing countries are still fighting the disease. TB is still evading them. How? TB can take naps. If TB is snoozing, it is called latent TB. In latent TB, the bacteria are gobbled up by macrophages—hangry microbe-eating champs. But TB survives the macrophage's wrath. The macrophage attracts neutrophils to the scene, which form a ring around the macrophage. This creates a fortress for TB called a granuloma. Inside, the bacteria doze away.

When TB is "procrastinating," with a trio of antibiotics and three months of healing (three times makes the charm, huh?), TB patients will be as good as new. Often, people infected by TB never develop TB disease. However, things can easily go wrong. Sometimes, TB bacteria will ignore the temptation of taking a nice, cozy doze and immediately clobber the body or stay undetected for years—maybe decades—before emerging when the infected person develops cancer or weakened immunity.

But the real clincher is TB's knack for developing antibiotic resistance. When a person stops using antibiotics before TB is fully KO'd or only takes one medication instead of several, TB bounces back, upgraded with antimicrobial resistance. Worse,

TB can develop resistance against *multiple* drugs. Treatment for drug resistant (MDR) TB has to be done very carefully. The drugs used to combat MDR TB are toxic and less effective than typical TB medications. These risky antibiotics also cause serious side effects, such as neurological damage, hearing loss, and kidney damage.

In industrialized countries, thirty-three percent of TB cases have the potential to develop antibiotic resistance, which is alarmingly high. But in impoverished countries, a mind-boggling seventy percent of TB infections are likely to be resistant to drugs. To top it off, a 2020 study revealed only one in three patients was given treatment for MDR TB! People living in crowded areas such as homeless shelters and prisons are also at high risk for MDR TB infections.

Rising from the Grave

TB's sneak attacks are also hard to detect. Like the average cold or flu, TB spreads when mucus containing the bacteria lands on someone else. But back in the 1800s, microbiology was still pretty much a black box. Believe it or not, people thought TB was passed down through genetics. However, this isn't the wackiest myth about TB. Old-timers believed people who died from TB rose from the grave and emerged again as vampires. Sounds like some sort of medieval post-apocalyptic horror story. People clamored to have TB corpses dug up and conducted rituals on these alleged vampires so they wouldn't haunt the living. Good thing microbiologists proved the public wrong!

Sniffing out the Enemy

Fortunately, us folks in the techy twenty-first century can diagnose TB. People living in Latin America, Africa, Asia, and Eastern Europe should be tested frequently. Folks spending time with TB patients or working at homeless shelters and nursing homes are also at risk. Doctors and nurses who take care of TB patients also need to be tested often. Individuals susceptible to TB infections are the elderly, infants, drug addicts, and people with medical complications. A skin or blood test can sniff out TB, but the bacteria can evade the tests if a person has a weakened immune system or was infected recently. If someone has taken the TB vaccine, the test may show a false positive. TB is sneaky; a positive test doesn't tell if the TB is active or inactive. For a proper diagnosis, people need a chest X-ray, sputum culture, or PCR test. Some doctors might have to take samples of body tissue to see if anything fishy is going on.

The Road to Recovery

It is very hard to recover from TB. Even with the best medical treatment, TB can relapse. Sometimes, the bacteria can infect the head and spinal cord, creating a fatal condition called TB meningitis. Sometimes the germs pop up in the bone marrow, the bladder, the joints, and the digestive tract. TB is very beefy; the strongest TB microbes can harass the immune system for a long time. It is common for people treated for TB to bear permanent lung damage. Diabetics with latent TB infections are more likely to develop the actual disease compared to nondiabetics. If untreated, the combination of diabetes and TB can be fatal. However, people who have HIV face bigger problems. When HIV enters the body, it weakens the immune system. With wimpy white blood cells, TB unleashes a full onslaught on the body.

Fortunately, there are little things you can do to reduce TB infections. Good ventilation is crucial to prevent TB from being inhaled; the bacteria are incredible acrobats and stay suspended for hours in the air. Sunshine can fry the bacteria. Cover your mouth when you cough or sneeze. Be sure to stay home when you are very sick. Come on, at least you get to skip school! But at the end of the day, early treatment is still the key to preventing TB from spreading.

There is also World TB Day. Every year on March 24—when Dr. Robert Kotch discovered TB bacteria—people recognize TB patients. This day raises awareness about the disease. After all, the BCG vaccine hasn't been updated in more than a century. It's working: many pharma companies are sprucing up improved TB vaccines. With new advances in medicine, TB can hopefully be eradicated in the future!

CHAPTER 15

FOOD BLUES

WHAT DO CHICKENS, COOKIE DOUGH, AND VEGETABLES HAVE IN common? A chance of being contaminated with *Salmonella*. Salmonellosis (commonly referred to as food poisoning) is a common illness in the US. Ironically, I caught this stomach bug when writing this chapter from eating undercooked chicken! There are 2,600 serotypes—or unique varieties—of *Salmonella*. This hodgepodge of bacteria is sorted into three groups based on pathogenicity.

Generalized *Salmonella* causes diarrhea, stomachaches, and a fever—classic stomach bug symptoms. Generalized *Salmonella* aren't picky about their host and attack a whole zoo's worth of animals, from insects to mammals. Fortunately, this gang of microbial desperados is relatively harmless. Generalized *Salmonella* aren't specialized to infect any certain host, putting an upper limit on their killing prowess. More severe illnesses are caused by host-adapted *Salmonella*. These are specialized strains that infect specific hosts, such as sheep, pigs, chickens, and humans. Fatal infections are caused by host-restricted *Salmonella*, which invest all their resources to take down *us*, good ol' *Homo sapiens*. Gulp.

Hoodwinking the Big Eaters

Salmonella invades the body when you chow down on contaminated food. Usually, your stomach is a death pit for bacteria; it has a pH of two, which is as acidic as battery acid. But when you fill your stomach with cheeseburgers, its pH rises to six, similar to the acidity of water. This allows *Salmonella* to safely frolic around in the stomach after you've gorged yourself. Even though intestines are covered in friendly microbes known as gut flora, the invaders still have a chance to infiltrate your body. *Salmonella* thrives in the mucus of the gut, and sometimes even goes en route to the bloodstream. Although *Salmonella* bacteremia (bloodstream infections) only occurs five percent of the time, these infections are deadly if antibiotics don't come to the rescue in time. Bacteremia or not, *Salmonella* is one of the leading killers of young children. While diarrhea is a frequent subject of toilet humor, severe cases starve the body of precious water and nutrients, especially dangerous for malnourished kids.

Salmonella has other vile tricks up its sleeve. Beefy white blood cells called macrophages are the power force of the immune system; they stay absolutely unopposed as they munch their way through microscopic battlegrounds. However, *Salmonella,* in the words of scientist Mayuri Gogoi, "hoodwinks the big eater to prosper." Macrophages digest their victims with endosomes—cellular stomachs filled with destructive enzymes and acid. Think bottomless pits in Super Mario levels. But when macrophages gobble down *Salmonella*, the bacteria are stuck in a special endosome called a SCV, or **S**almonella **C**ontaining **V**acuole. When the SCV oozes acid, *Salmonella*'s cytoplasm (the gooey jelly inside a bacteria) turns more acidic to neutralize the harmful effects of the acid. *Salmonella* also quickly adjusts to the low-nutrient conditions inside the macrophage. When time is right, *Salmonella* kills the macrophage and bursts out!

Salmonella also wields clusters of genes called **S**almonella **P**athogenicity **I**slands (SPIs). These "islands" are no tropical paradise; SPIs are honed for combat. SPI-1 is used to murder macrophages. *Salmonella* hides in macrophages by activating SPI-2. SPI-4 produces endotoxins and exotoxins. Exotoxins are toxins secreted when *Salmonella* is still alive and kicking. Endotoxins, on the other hand, are released when the bacteria are killed in a James Bond-ish final onslaught. Lastly, SPI-6 produces (or deactivates) proteins to mess around with your immune system. Mini DNA upgrades called plasmids also enhance *Salmonella's* infectious power.

But don't worry! Your body is home to an army of white blood cells, ready for action. With your faithful fighting force, doctors usually let *Salmonella* infections run their course without the help of antibiotics. Even with some of their macrophages down, many people with salmonellosis think they've caught the flu. However, even when the bacteria are kicked out from the body, it causes nasty bouts of constipation and diarrhea for weeks on end, even in brawny, healthy youngsters. Gross.

Named after the Scientist, not the Fish

Before we go further, let's talk about some history. Back in the 1800s, people didn't know anything about *Salmonella* until Doctor Karl Joseph Eberth discovered the bacteria grubbing around inside his patients in 1880. In his honor, *Salmonella* was originally named *Eberthella!* Bacteriologist Georg Gaffky successfully isolated the bacteria four years later. Dr. Daniel Elmer Salmon, a veterinarian, and his assistant, Theobald Smith, discovered the first strain of *Salmonella* in 1885. However, Dr. Salmon rushed down the wrong road by claiming *Salmonella* was the culprit of hog cholera (aka classical swine fever), a deadly viral infection in pigs. Despite Dr. Salmon's mistake, bacteriologist Joseph Lignières later renamed the bacteria *Salmonella* in his honor, not after the fish. (Coincidentally, salmon carry *Salmonella* too.)

A "Typhoid" Lethal Bacteria

With macrophage-busting abilities, it's no surprise some strains of *Salmonella* are lethal. *Salmonella typhi* is the most infamous member of the *Salmonella* horde. Its cousin, *S. paratyphi*, is less common and severe but still sparks unpleasant and deadly infections. Also known as typhoid and paratyphoid fever, these bacteria ravage the body with rosy skin rashes, appetite loss, fits of coughing, constipation, and persistent fevers. Fortunately, most cases of typhoid/paratyphoid fever in industrialized countries are caught while traveling. Inside well-sanitized regions, you almost never catch typhoid fever. Each year, the US reports around a measly 350 cases of typhoid fever and ninety cases of paratyphoid fever. However, outside these safe havens, there are plenty of opportunities for *Salmonella* to strike. For every person confirmed to have a *Salmonella* infection, thirty more are also infected. The true number of people infected by *Salmonella* is vastly underestimated.

There's a vaccine to combat *Salmonella*. The vaccine is gulped down inside a pill or injected into the muscles. It doesn't fully prevent *Salmonella*, but it makes infections a lot less severe. The vaccine prevents this foodborne fiend fifty to eighty percent of the time; the reason why we don't use it frequently is because the vaccine occasionally causes a nasty slew of side effects (e.g., swelling, hives, difficulty breathing/swallowing, and even fainting).

Rough and Ready

Salmonella is a pretty versatile bacterium. Most species move on their own thanks to their tail-like flagellum, allowing these bacteria to scoot away from danger and to find cozy nooks. *Salmonella* can take on temperatures from 41 to 117 degrees Fahrenheit, though especially hardy bacteria take on temps of 36 to 129 degrees Fahrenheit. But everything—even bacteria—has a comfort zone. *Salmonella* likes ninety to ninety-five degrees best, explaining why this guy vacations in the tropics. *Salmonella* needs to guzzle lots of water to multiply, but it can haul itself out of the wet, swampy areas where it lives and accommodates to a drier life. Bet these fellas would like some lotion though.

Don't Cuddle Lizards

There have been tons of *Salmonella* outbreaks. One was traced to bearded dragons (not actual dragons). There were fifty-six reported cases, and nineteen patients landed in the hospital. Twenty-six US states were involved in the outbreak; fortunately, no one was killed. Another outbreak occurred in the backyard poultry. There were 562 cases reported, along with ninety-two hospitalizations and two deaths. The mega-outbreak occurred in all forty-eight mainland states. However, the total number of cases in both outbreaks is much higher; most people don't test for *Salmonella* and get better on their own.

These outbreaks occurred because unwary people snuggled/kissed their pets or let them loose in the house, contaminating their food. (Personally, I'm not so sure if lizards are the cuddly type.) To prevent catching *Salmonella* from your scaly and feathered friends, spiff up and disinfect their homes to avoid *Salmonella* germs dwelling inside. Also, wash your hands thoroughly after frolicking with your pets!

Another *Salmonella* outbreak was caused by a contaminated batch of peanut butter. This peanut butter was sold worldwide; thankfully, only sixteen people were infected and two were hospitalized. The FDA tracked down all jars of contaminated peanut butter and urged people that munched on contaminated chows to disinfect surfaces these products have touched and throw away these foods. Be cautious around peanut butter!

Typhoid Mary

In ye old times, people knew about *Salmonella* but had no way to treat the disease. Doctors diagnosed this stomach bug based on patients' symptoms and quarantined them until they recovered. One particularly famous *Salmonella* patient was Typhoid Mary, formally known as Mary Mallon. Mary claimed she was born in the US, but she was actually from Cookstown, Ireland. Mary immigrated to New York City in 1883, where she worked as a cooking servant traveling from house to house. However, she asymptomatically carried *Salmonella typhi* around with her. Even with no sign of the disease, Mary accidentally spread the bacteria through her cooking. Unfortunately, whenever she was suspected of causing a typhoid outbreak, she moved on to the next house!

This continued for a while...a long while. From 1900 to 1907, mysterious outbreaks swept through New York. Eventually, NYC's sanitary engineer, George Soper, was called to investigate the strange scenario. He believed the outbreaks were caused by Mary, especially after a particularly deadly outbreak in Manhattan in 1907. After Soper met with Mary, he traced all twenty-two cases of *Salmonella* in that outbreak to her. Mary was soon on the run, hotly pursued by a whole cast of authorities, including Soper. Eventually, she was sent to an isolation center in North Brother Island: a gloomy, secluded place that is—somewhat ironically—a bird sanctuary.

She stayed on North Brother Island for three long years and was released in 1910. Mary was strictly prohibited from being a chef ever again. Despite this, she *still* worked as a cook...in a hospital! Soper started to sniff her out again after two deadly outbreaks in New Jersey and New York. He finally found her in a suburban home and took her to North Brother Island, again. She died in 1932 due to a stroke after decades of struggling between fighting for personal freedom and being a public health threat.

Salmonella has put many famous people in jeopardy, not just Mary. Orville Wright, one of the famous Wright Brothers, contracted *Salmonella* in 1896. His brother Wilbur Wright nursed him back to health for six weeks. If he had died, airplanes probably wouldn't exist! But in a cruel twist of fate, Wilbur died at the age of forty-five from *S. typhi*.

Rise of a Superbug

Salmonella is a superbug. You are probably familiar with antibiotic-resistant microbes after reading seventy-five percent of this book. Remember those nifty plasmids I talked about a few pages ago? Those little guys are *Salmonella's* secret in becoming a superbug: some plasmids are equipped with *eight* drug-resistant genes! How did this happen? In the 1900s, antibiotics were considered a miracle cure, especially by the livestock industry. Antibiotics were used to promote growth and protect farmers' valuable livestock from disease. This abuse caused bacteria like *Salmonella* to develop antibiotic resistance. During the 1960s, most farm animals miserably hunched in corners fighting nasty stomach bugs. Then came the "aha" moment. It was antibiotic resistance!

Salmonella-related antibiotic resistance has spread like wildfire, especially in Africa. In Uganda, *Salmonella* is a nuisance to the egg industry. In Nigeria, quail farming is becoming a hassle. A few hundred miles away in the Middle East, a wonder drug used to treat severe salmonellosis called ceftriaxone is rendered obsolete. But all this pales to what is happening in Western Asia. Countless drugs equipped with life-saving capabilities like cephalosporins and fluoroquinolones (new tongue twisters of the day) are painfully ineffective.

However, there is a tiny glimmer of hope. Some drugs still work against *Salmonella*. Organizations tracking *Salmonella's* drug resistance have a whole Pokédex-style catalog of drug-resistant *Salmonella*, potentially useful for creating new antibiotics. Outside the big-brain medical industry, farmers discourage the use of antibiotics unless absolutely necessary lest further resistance emerges. Hopefully, this will be enough to fight back and turn the tide against *Salmonella*.

A Galling Affair

Salmonella also causes gallbladder cancer. The gallbladder produces bile, a greenish-yellow liquid that breaks down fats and proteins. After being walloped by the immune system, some strains of *Salmonella* retreat into the gallbladder and stay unnoticed for years. This can cause bile to harden into gallstones. Gallstones stuck in the bile duct trap fluid inside the gallbladder, increasing the risk of deadly gallbladder infections. If these pesky stones are jammed in the gallbladder themselves, they tamper with liver function. If the liver doesn't filter out waste from the bloodstream, the skin and eyes turn a sickly yellow. This is known as jaundice. To top it off, gallstones also irritate the pancreas. Cells exposed to abnormal amounts of bile develop into cancer. Yikes!

Get Out!

Although *Salmonella* hitchhikes on foods to do its dirty work on humans, other foods can combat this bacteria's symptoms. When faced with *Salmonella*, avoiding diarrhea medications is the first step. It's important for your body to get the bacteria out of your system. Diarrhea medications trap *Salmonella* in your digestive tract, which is *bad*. Instead, sip a cup of ginger tea, which soothes the stomach but won't trap *Salmonella* inside. The BRAT diet also prevents you from losing your lunch. BRAT stands for **B**ananas, **R**ice, **A**pplesauce, and **T**oast. These foods don't have particularly strong flavors so they don't induce vomiting, helping your grub stay in your body. Similar foods include Jell-O, potatoes, oatmeal, honey, egg whites, and crackers. Avoid foods that are hard to digest, such as fatty, tough, acidic, and/or spicy foods. (We're looking at you, buffalo wings.)

Now, let's look at my *Salmonella* experience to see what you can do if you are run down by stomach trouble. First, don't look at the computer. Even though this means not checking Reddit for a few days, computers make nausea even worse. Fortunately, watching TV is fine. Don't be afraid to laze on your bed and take a nap. Sleeping on your stomach is discouraged; this makes stomach cramps worse. Drink plenty of water to make sure you are hydrated, but don't chug. Though every *Salmonella* infection is different, the bacteria only cause serious agony for only one day. Only a stomach-ache is left on day two, and you (hopefully) should've recovered by the third.

And Stop Eating Raw Cookie Dough

Salmonella seems like a trivial threat. But this bacterium is very common, and there is strength in numbers. Many people around the world aren't healthy enough to fully fight off the infection and suffer for weeks or even months. Worse, the threat of antibiotic resistance looms. But if we work together—both in hospitals and communities—to use antibiotics more wisely, we shall succeed. It won't be easy, but if we are determined, we will win the fight against *Salmonella*. And please, please, stop eating raw cookie dough.

CHAPTER 16

DEATH BY SLOBBERING DOGS

YOU CAN DIE FROM SLOBBERING DOGS, OR MORE APPROPRIATELY, rabid dogs. Excessive drooling is a hallmark of dogs overcome by rabies, but rabid skunks, foxes, racoons, seals, and generally any other mammal infected with rabies (including humans) also slobber. In the movies, rabid animals are easy to identify: they are aggressive, foaming at the mouth, and drooling enough saliva to fill a swimming pool. In 1932, Disney released a film called *The Mad Dog*. When Pluto—Mickey Mouse's beloved dog—accidentally swallows a bar of soap, bubbles cascade out of his mouth and he is considered rabid. A flabby, bad-tempered cat dubbed dog-catcher Pete pursues Pluto, and the chase goes on! The cartoons, however, heavily understate the dangers of rabies.

Slow and Steady Wins the Race

Rabies is *no joke*. Rabid animals can be aggressive slobber mouths, but others are tame, meek, and even act unnaturally cute. Some animals might even look completely okay; rabies hides in their nervous system for days, weeks, months, and even *years*. Each day, rabies slowly crawls twelve to fourteen millimeters (0.4 to 0.5 inches) up their nerves. When an unfortunate person is bitten in the arm, they have to rush to the ER for a vaccine much faster than someone bitten in the leg. The longest confirmed incubation period for rabies was seven years, and some scientists report incubation periods of a whopping *twenty-five years*. But the reasons behind the virus's long slumber are unknown.

This slow incubation allows rabies to stealthily hijack the body, but it can also expose the virus's weakness: vaccination. Rabies is so slow that people can receive the vaccine after they're infected! This form of vaccination is called post-exposure prophylaxis. It has saved thousands of lives around the world. Rabies also has another weakness: it can't survive for long without the warmth or saliva of an animal. Outside the safety of a body, rabies only lives for several seconds. However, the virus can keep a toehold in dead animals for up to two days; one of the many reasons why you shouldn't touch roadkill!

There is a common urban myth that opossums carry rabies. These gray, furry marsupials have body temperatures too cold to harbor rabies. By the way, opossums are some of the toughest animals on the planet. They are immune to botulism toxin, snake venom, honeybee stings, Lyme disease, and scorpion stings. Other small mammals (e.g., rodents) generally don't carry the virus either. Mice and rats are literally furry popcorn to larger animals. It only takes one swift bite from an animal, rabid or not, to kill them, mercifully saving them from the slow death of rabies. Birds, reptiles, insects, and fish are also free from rabies's horrors.

Though rabies is almost eradicated in the US and Canada, it is still common across the globe. Someone dies from rabies every ten minutes. This alarming death toll is the result of the scarcity or unavailability of the vaccine in these areas. Even the most powerful ancient civilizations were at the mercy of rabies without a vaccine. Rabies was first described more than 5,000 years ago in Mesopotamia, a mighty kingdom in the Middle East. The Greeks also wrote about rabies. Homer once described one of his characters in *The Iliad* as rabid when they became berserk during a battle. Lyssa, the goddess of wolfish rage, allegedly possessed opposing warriors Hector and Achilles. In fact, rabies is also known as lyssavirus. Lyssa was also depicted turning dogs rabid to attack a hunter who was turned into a deer for bragging his skill was better than Artemis, the Greek god of the hunt. The Greeks also believed the dog star, Sirius, descended to Earth each summer, bringing the blistering "dog days" of summer as well as rabies outbreaks. But despite the telltale signs of rabies, people couldn't do much about it.

Hacking the System

Normally, the brain-blood barrier, or the BBB, blocks out harmful organisms and other nasty ones that would otherwise enter your head. It is made of a variety of cells—some even mimic white blood cells and destroy germs that enter the brain—and specialized blood vessels. If something busts a gap in the BBB and squiggles into the brain, the BBB triggers an inflammatory response that creates holes in the BBB, allowing white blood cells to pour in. This inflammation is a double-edged sword: it fights off infection but potentially causes brain damage. Rabies, on the other hand, completely escapes the response. The chemicals released to dig holes into the BBB are not released when the virus squirms into the brain, making the immune response futile.

Rabies also tricks the immune system into thinking it isn't there. The muscles' genetic material binds to the virus's RNA, keeping the nerves intact and the progress of the virus slow. This is why someone can drive safely to a hospital and get their rabies shot without their nerves crumbling. However, this also gives rabies a free ticket to the brain. The virus initiates apoptosis (programmed cell death) in white blood cells called T-cells, which analyze pathogens and call for backup. This makes rabies like a ninja stalking the nerves and sneaking up to the brain, careful to leave no trace...until it's too late.

Rabies hotwires the brain, causing vital functions you take for granted (e.g., drooling, digestion, and heartbeat) to be messed up. Some patients literally drown in their own spit. If you think that's funny, *it's definitely not*. Rabies is no laughing matter; people suffering from the virus deserve all the sympathy and hope they can get. In fact, half of the patients die of heart arrhythmia—when the heart beats too fast, too slow, or too irregularly. (Note: it is still normal for your heart to change pace when you're running and sleeping.) Hallucinations, double vision, and other nightmarish symptoms occur as well. By the time the immune system responds, all is lost.

But there is always a chance some people can survive. Jenna Giese, bitten by a rabid bat when she was fifteen, survived rabies by pure luck. How? Rabies scrambles the brain and causes its fuses to blow. However, when put under a medically induced coma, people can survive because the brain "hibernates," slowing down the effects of rabies and allowing immune cells to come. This approach is called the Milwaukee Protocol. With the aid of antivirals, Giese managed to stagger back to recovery. She lost her ability to talk and walk, but she bounced back, and Giese went on to graduate from college and start a family.

However, Giese was extraordinarily lucky. Others have undergone the Milwaukee Protocol and died. The ethics of it remain volatile: the Milwaukee Protocol has saved lives, but it comes with huge drawbacks. The process *itself* can kill with shock, arrhythmia, and edema. Medical researchers are racing to find alternative treatments after rabies enters the brain, but the clock is ticking. Some researchers claim Giese's recovery was due to the infection with a rarer, less lethal form of rabies. Tracking how the virus gets into the brain and reversing the damage are ideas on the drawing board, but the way forward remains unseen.

Despite the dangers of rabies, it is 100 percent preventable. The rabies shot has two parts. Part one—Human Rabies Immune Globulin (HRIG)—is made from antibodies from people already immune to rabies. HRIG provides a brief flood of antibodies to hinder rabies before the body creates its own antibodies to finish the virus off. It's usually injected in/around a bite wound. However, dosage must be carefully administered; otherwise, HRIG might suppress the body's own antibody production. Nevertheless, HRIG ensures part two of the shot—the actual vaccine made from killed rabies viruses—will come to the rescue in time. The vaccine is given in several doses on days zero, three, seven, and fourteen. Depending on the vaccine, immunity can last as long as ten years! You might've heard getting the rabies shot involves being impaled with seventeen needles into your stomach. While recipients of the vaccines up until the groovy '50s needed to contend with being a living pincushion, the four to five shots you will receive today will be relatively painless. Despite the efficacy of vaccines, all wounds—whether they are a bite from a rabid animal, a nasty cut from tumbling off your skateboard, or just simply a skinned knee—need to be washed/cleaned immediately to reduce the risk of contracting an infection.

Going Batty

Bats, in particular, are notorious for spreading rabies. They may take all the blame, but bats are actually innocent. These fluffy fliers are curious, intelligent animals that live colorful, diverse lives.

A reason why people fear bats so much is concern about contracting disease. Despite living on wholesome diets of nectar, fish, and fruit, bats have immune systems as tough as nails. It's hard to tell if these airborne fuzzballs are infected or not. Thankfully, healthy populations of bats don't harbor much rabies. In fact, bats actually *prevent* disease by munching on mosquitos.

Rabies makes infected bats act like angry, haywire gliders; however, it isn't the virus's fault. Rabies just wants to reproduce and create the next generation of baby viruses. But for a bat (or any other animal) to have the potential to spread rabies, it has to either bite or lick a wound (no matter how small). If you see a suspicious wound on your body when you wake up or see bats fluttering around the house, flying during the day, lying around on the sidewalk, or showing unusual behaviors, call the wildlife control center and the doctor, pronto!

Sometimes, people don't make the connection between rabies and a bat bite because bats have dainty, needle-shaped teeth. These pearly whites hurt, but they don't leave much of a mark. On the other hand, vampire bats intentionally cut through you to slurp up some blood. Of the 1,300 species of bats, the vampire bat is the only one that actually drinks blood. This little fella rarely attacks humans, but thirsts for cattle blood. Vampire bat saliva prevents blood from clotting, so blood continuously flows to create a disgusting protein shake for the bat. The vampire bat doesn't hurt its victims, but these bloodthirsty mammals frequently spread rabies among cattle herds. You shouldn't worry about giving up your cheeseburgers; rabies is easily destroyed by cooking. However, rabies is a big pain in the neck for farmers. Infected cattle often die before they provide enough meat and milk. They drool excessively and struggle to swallow their cud, starving them to a foul, frothy death.

If you're starting to get queasy from these bats, you aren't alone. The ancient Mayans cowered when a massive vampire bat named *Desmodus draculae* flew overhead. This beefy bat might've even inspired the creation of the Mayan bat spirit Camazotz. But you don't need to worry about these behemoths feasting on your blood. What's more pressing is that vampire bat habitat is disappearing, forcing bats to look for new homes. Farms, abandoned buildings, and wells are just as cozy as the rainforest canopy but bring bats close to humans, increasing the likelihood of them feasting on human blood. Climate change also drives vampire bats further and further north from the scorching equator. Experts are worrying these bats will enter the US and spread new variants of rabies. Poison and other attempts keeping vampire bat populations at bay have failed. Welp, nature strikes back! So, be nice to bats...and keep your distance.

Wrangling with Rabies

One sunny day back in 2017, two Florida residents enjoyed a delicious salad of leafy greens. But the next day, they found a *bat carcass* in their salad. Rabies can't survive on animals for long, but both people were quickly tested and vaccinated for rabies. The uproar died down, but this was not the first time some exotic meat was added to a salad. Other people have also plucked dead birds and other small animals from their salad—a gruesome result of mass harvesting. Watch out for these extra bits of unexpected protein supplements!

Having an hors d'oeuvre of a bat with your salad is rare, but there's nothing you can do to avoid critters diving into your salad. On the other hand, you can easily prevent animal-spread rabies. If you have a dog or cat, make sure to take Fido or Ms. Whiskers to the vet to get vaccinated, just like many other pets around the world. Never touch wild animals, dead or alive, even if they're friend shaped. And *never* hesitate to tell someone if you're bitten by an animal. If you follow these basics, you're pretty much set against rabies. But due to the virus's prevalence, some people take bigger measures.

People across the globe are giving wild animals oral rabies vaccines concealed in tasty baits sprinkled down from helicopters. This bait is designed to appeal to raccoons—common neighborhood crooks. These gourmet baits are topped with an alluring drizzle of fishmeal. When racoon gourmands sink their teeth into these exquisitely prepared baits, a life-saving vaccine slides down into their stomach. Other appealing varieties include blocks made of (digestible) plastic and fishmeal and packets sweetened with icing sugar. Different appetizing designs are specially made for gray wolves, coyotes, and other animals, making these trendy treats available for everyone. Most of the bait is consumed after five days. Carnivores, scramble into the forest now. These treats are limited edition!

However, some folks don't have this luxury. People living in Haiti have a particularly large rabies problem, especially after suffering from a devastating earthquake in 2010. Haiti has less than forty-five percent of its dogs vaccinated. Fortunately, determined people with a dash of creativity are ready to help. An app tracks vaccine campaigns in Haiti. Increased sponsorship of the Integrated Bite Case Management (IBCM) led to a massive rise in rabies vaccination and a huge drop in rabies deaths. 20,000 comic books featuring popular cartoon character Ti Joel were sent to schools all around the country. All of this paid off: in 2019, Haiti only suffered three human deaths from rabies, compared to thirteen deaths in 2011.

Ethiopia is also wrangling with rabies, a burden to its agriculture-based economy. Intriguingly, Ethiopians gently catch dogs with nets to vaccinate them. Dog catchers are shown slow-motion videos for a demo. Fortunately, brute force isn't always needed: street dogs have strong friendships with people and are easy to vaccinate. Rabies vaccination committee members have also consulted the Ethiopian government for help. Like in Haiti, Ethiopians are gaining ground against this virus!

We're Almost There!

Rabies is getting pushed to the brink, but some countries need just one more push to help eliminate the virus. Cooperation and innovation will help us achieve this goal. We are almost there, but we need a final boost. Unlike so many other brawls with pathogens, the fight against rabies will likely have a happy ending. Who knows what will happen next in this battle? A new drug? A new vaccine? Or maybe even a doctor that will revolutionize medicine. Maybe that will be you!

CHAPTER 17

VAMPIRES IN ADOBES

THERE ARE A LOT OF HOUSES. FORTS MADE FROM BLANKETS THAT allow you to evade devastating pillow fights. Huts squashed together so they touch each other's walls. Shacks so old and shabby you want to dial a demolition crew to topple down the ugly spectacle. Cozy igloos made from fluffy blocks of snow. But microbes have houses too. Algae called diatoms live in tiny glass houses that look like jewelry. Some bacteria drift around in fluffy clouds—the ultimate carpeted home. However, not all the houses microbes dwell in are quite so scenic!

One of these homes is odd and unappealing. *Trypanosoma cruzi* is a squiggly little sporozoan protozoa that looks like a banana with a tail. But don't let its cartoonish appearance fool you! *T. cruzi* lives in *Triatomine* bugs—pesky little critters with needle-like mouthparts and a thirst for blood—hence the name of "vampire bugs." Because these bugs bite the face, they are also dubbed the overly romantic name of "kissing bugs." *Triatomine bugs* are also called "conenose bugs," "benchuca," "vinchuca," "chinche," and "barbeiro." However, these bugs are only *T. cruzi's* mobile homes and RVs. What these parasites desire is a state-of-the-art human heart!

Crawling Right In

Also known as Chagas disease, *T. cruzi* was named after its discoverer, Carlos Chagas. The cunning parasite uses the penny-sized kissing bug as a vector, an organism where it can multiply and hitchhike to another host. After kissing bugs gorge themselves on blood, they defecate near the wound. These bug plops contain millions of parasites. The bite soon gets itchy, and the urge to scratch it is irresistible. Scratching breaks the skin and *T. cruzi* can sneak into the bloodstream and drift en route to the heart. People living in South and Central America are at high risk of acquiring Chagas, but other Americans are also vulnerable. In rural parts of the Americas, houses are often constructed out of mud bricks (known to some as "adobe architecture") and/or have thatched roofs made of plant matter. Kissing bugs love hanging out here, so they crawl right on in.

In more urban areas, potential kissing bug real estate includes cracks in the sidewalk, doghouses, chicken coops, leaf piles, tree bark, small nooks and crannies in rocky structures, porches, rodent nests, your mattress, or any other structure close to your bed. Kissing bugs will live in any place near a potential blood source—whether that be humans, birds, reptiles, or dogs—and emerge at night to feed. These insects unknowingly carry their parasitic cargo of *T. cruzi* with them, infecting humans (and our pooches) while we sleep. Kissing bugs can't be caught as easily as cockroaches or ants, either; roach hotels or ant baits are not effective against kissing bugs.

T. cruzi has a ton of tricks up its sleeve. Besides being spread by kissing bugs, the parasite can also be transmitted to unborn babies by their infected mother. Babies are often not diagnosed with Chagas because they carry low parasite loads; worse, the tests used to diagnose Chagas disease are faulty. Many babies are asymptomatic at birth. To top it off, the chance of Chagas disease being transmitted in utero is a measly one to five percent.

This sounds like good news, but it means most healthcare providers are unaware of Chagas. Risk factors for this transmission are not fully understood, but immunocompromised mothers and/or mothers with high parasite loads are more likely to pass the parasite to their baby.

Patients also contract *T. cruzi* through contaminated heart transplants or blood transfusions. Both donors and physicians possess limited knowledge about Chagas. Though one in 27,500 blood donors potentially has *T. cruzi*, this adds up, considering seven million blood donors donate 13.6 million blood transfusions annually. Fortunately, blood drives provide routine screening for Chagas disease, hopefully curbing *T. cruzi* from hitchhiking on donated blood. Eating undercooked rabbits and armadillos also spreads Chagas. Rabbits and armadillos are reservoirs for *T. cruzi*; these furballs are infected but don't get sick themselves.

From the Bottom of My Heart

When *T. cruzi* intrudes someone's body, swarms of white blood cells such as neutrophils and macrophages come out to engage the abominable protist. Antibodies are hurled everywhere. Flu-like symptoms occur. This is the acute phase of Chagas; this is when most invading parasites are killed. For some lucky people, this is the end of the story. Unfortunately, the surviving parasites often slink away and go straight for the heart. However, severe symptoms only pop-up decades later. Nestled deep inside the body, the parasite diverges into a variety of different forms. Though these Pokémon-esque pathogens all have goofy names, they mean business. Metacyclic trypomastigotes are the first to engage the immune system. Epimastigotes—the reproductive form—invade the gut. Circulating trypomastigotes drift around in the bloodstream. Trypomastigotes "evolve" into amastigotes, which burrow into the heart. Amastigotes lose the ability to swim and become dormant, replicating slowly. However, all these forms share a trademark kinetoplast—a mass of DNA near the nucleus.

As the amastigotes wriggle deeper into the body, they act like ticking time bombs. Twenty-five to thirty years after infection, twenty to thirty percent of infected people begin to display concerning symptoms—heralding the chronic phase of Chagas. *T. cruzi* slowly but steadily tears holes in the heart, causing chronic inflammation and tissue damage. Bit by bit, the parasites literally eat the heart from the inside out!

The heart, once a yoked-up organ, suffers from blockages and myocardial insufficiency. Myocardial insufficiency occurs when the heart struggles to beat due to muscle damage. Not good! Another nasty heart disease is called left-sided heart failure. This doesn't mean the left side of the heart has gone kaput, but it does come pretty close. The heart has four chambers known as ventricles. Ventricles open, close, and flex to pump blood. These little fellas make your heartbeat. If the left ventricles are damaged in left-sided heart failure, erratic heartbeats, fatigue, chest pain, swelling, and shortness of breath occur. If *T. cruzi* isn't knocked out by then, right-sided heart failure occurs. This condition is often a result of left-sided heart failure, but can appear on its own. The symptoms are similar but more severe.

If medical attention doesn't save the day, the king of ventricle-killers arrives: biventricular heart failure. This means your heart is functionally *dead*. When worst comes to worst, a heart transplant is a last resort. But why don't patients get heart transplants right away? First of all, heart transplants cost a lot of moolah. Second, successful transplants require the use of immunosuppressants—drugs that weaken the immune system. Without taking immunosuppressants, patients run the risk of organ rejection. Organ rejection happens when the body's immune system attacks a donated organ, mistakenly thinking it's a foreign "invader." Rejected hearts suffer from failure and heart attacks. But immunosuppressants allow scattered bands of *T. cruzi* to reactivate and take over weakened bodies once more. Antiparasitics and close monitoring are a must after a heart transplant!

T. cruzi doesn't just convert the heart into its stomping grounds. It renovates the entire body into a massive, infested mansion! It uses saliva-producing glands as swimming pools and slides down the esophagus as if it were a water slide. This widespread esophagus infection can lead to a painful condition called megaesophagus, which results in nasty inflammation, trouble swallowing food, and wheezing and coughing (especially if the affected victims are lying down). Worse, megaesophagus jams the airways with food crumbs and saliva by squeezing the trachea. The stomach and intestines also serve as hallways and corridors for *T. cruzi*. There, it starts bouts of megacolon. Megacolon is a medical term for an inflamed intestine. It causes horrible belly-aches and even *tears the intestine apart*. Eventually, sepsis gives the final blow!

Blame the Doggos

T. cruzi has been around ever since we have spelunked in caves. Most infected animals did just fine. However, *T. cruzi* often hassled wolves. When humans tamed these pups with treats and puppy love, *T. cruzi* stuck along for the ride. Eventually, the parasite "jumped ship" and wreaked havoc among us, too. This actually isn't all that unusual: pathogens constantly rove the nooks and crannies of our planet, always looking for new hosts to bring down.

We've painted a pretty bleak picture about Chagas so far. Fortunately, it isn't all doom and gloom: powerful methods of diagnosis and treatment easily bring *T. cruzi* onto its knees. During the acute phase, blood tests scan samples of blood to detect any infiltrating *T. cruzi*. Another warning sign that helps diagnose Chagas are chagomas (aka Romana's Sign)—itchy red bumps near the eyes. Since kissing bugs like to bite the face, they leave behind clues that betray their presence. If *T. cruzi* is found gnawing away the body, the parasites are given a full-force blow with antiparasitic drugs.

The chronic phase is harder to reckon with. Blood parasite exams are ineffective; most of the parasites are hidden in the heart or digestive tract. Antibody tests are more useful because the immune system still grapples with *T. cruzi* during the chronic phase. But antibody tests are inaccurate: they must be used multiple times for clarity. Other factors (e.g., where patients live) are also put into account. The antiparasitics used to treat Chagas disease (e.g., benznidazole and Lampit) have drawbacks too: side effects caused by the drugs increase with age. Patients under age fifty usually suffer no adverse symptoms, but elderly patients are tricky to treat. Doctors must carefully weigh the benefits of eliminating *T. cruzi* against the health risks of elderly patients. Drug companies warn benznidazole can increase your risk of getting cancer and even *mess with the DNA in your chromosomes!* (Or at least that's what studies on rats suggest.) However, there's a silver lining to Lampit and benznidazole: these drugs are commercially available, affordable, and conk out *T. cruzi* ninety percent of the time.

(Don't) Squash that Bug!

When faced with a kissing bug, don't swat or squash it. The parasites may squirm through tiny cuts in your hand. Instead, put the bug into a container filled with rubbing alcohol and take it to your local lab or health department for identification. *T. cruzi* lurking around your house can be killed with a solution made from one part bleach to nine parts water or seven parts alcohol to three parts water. If kissing bugs are prowling your home, call the pest control immediately!

Shockingly, most people infected with *T. cruzi* don't know they have the parasite swimming around in their blood! Ninety-nine percent of people dismiss Chagas disease as a cold and don't bother testing for the parasite (at least initially). However, people should stay on their toes. We need to inform more people about this silent killer. Many governments and committees are doing this now. With just a quick Google search, you'll see they're many organizations determined to stop *T. cruzi*. Chagas drugs are also double-edged swords. They cure the parasite but cause a nasty cocktail of side effects. The pharma industry really needs to spruce up itself when it comes to treating *T. cruzi!*

The tough work shouldn't stop here. After reading this chapter, *you* can also act. Donating to an organization devoted to fighting Chagas is a good first step forward. If you plan to think big, you can spread the word about this parasite by making a blog post or short message on social media. Or simply find resources that'll help you protect you and your loved ones from *T. cruzi*. We can all work together to stop this silent menace—of us. Watch out *T. cruzi*; people are hot on your tail! And oh, one more name for these troublesome *Triatoma* bugs before I go: assassin bugs. A much more appropriate name for these devious insects!

CHAPTER 18

ARMADILLO TROUBLE

ARMADILLOS ARE QUESTIONABLE CUISINE. BUT SOME PEOPLE EAT them anyway. These mini, biological tanks spread a variety of diseases, including the scary-sounding "leprosy." Leprosy—now known as Hansen's disease to protect human rights—is caused by the bacteria *Mycobacterium leprae*. Hansen's disease seems like a frightening, lethal bacteria, but due to natural immunity, ninety-five percent of all people who contract the disease don't suffer from any symptoms.

Contrary to what some people believe, Hansen's disease isn't very contagious. Once a person is cured, he or she is no longer contagious. Hansen's disease is spread by airborne droplets or by being in close contact with an infected person over a long period of time. Hugs, handshakes, and fist bumps don't spread leprosy. But this brings us over to armadillos. Nine-banded armadillos—native to states like Texas and Oklahoma—are the main reservoirs of Hansen's disease. The chance of getting leprosy from armadillos is rare; only twenty percent of the nine-banded armadillo population carries the bacteria. This adds up to a 0.01 percent chance of contracting Hansen's disease from an armadillo. But some people enjoy adding raw armadillo liver to ceviche—a mixed dish made from fish, shrimp, vegetables, and tortilla chips. Armadillo livers are potentially chock full of leprosy bacteria, but this doesn't stop some adventurous people from eating them anyway!

GRRR

Life in the Slow Lane

Hansen's disease is bizarre because it multiplies very slowly. People can display symptoms nine months after being infected, but symptoms usually pop up around five years later. Some people even show symptoms twenty years later! With this ample downtime, Hansen's disease can be easily knocked into oblivion with multidrug treatment. But if a patient doesn't receive treatment, they develop bumps and ulcers all over their body. These sores look painful, but infected people don't feel any pain: Hansen's disease damages the nerves. This nerve damage also causes a slow, creeping paralysis. Fingers, toes, and limbs freeze in place. Eyelids struggle to close. Before the mid-20th century, people were so afraid of Hansen's disease they claimed infected victims' limbs would rot and fall off. This is an urban myth, but people suffering from Hansen's disease sometimes need to amputate limbs due to severe inflammation.

Hansen's disease also causes further complications. *M. leprae* thrives in the iris (the colored part of the eye). This infection leads to blindness if left untreated. In severe cases of this dreadful disease, Hansen's disease even damages the kidneys. *M. leprae* changes its victims' lives forever. Besides eating away at the nerves, the bacteria also wear away muscle tissue. Even after receiving antibiotics, people suffer from extreme weakness. Their damaged nerves are unable to feel pain, making it easy for people to cut or burn themselves. Hansen's disease also causes skin to become wrinkled and deforms the nose. All of this cannot be reversed; this is why early treatment is crucial.

The Psychological Cost

The effects and injuries wrought by Hansen's disease are bad enough, but the psychological cost of the bacteria is even more damaging. The victims of Hansen's disease were and still are discriminated against and excluded from society. Patients were quarantined on tiny islands or on high mountaintops in places dubbed "leper colonies." The word "leper" is extremely hurtful and offensive to people suffering Hansen's disease and only brews more discrimination. According to WHO Goodwill Ambassador Yohei Sasakawa, this slur is "an extremely damaging term that contributes to stigma and impacts human rights." The preferred term to describe these places is "Hansen's disease colony." Oftentimes, Hansen's disease victims were banned from working, voting, or even seeing their loved ones. If a child were born in a colony, that child would be snatched away and raised elsewhere, far, far away. People feared leprosy was contagious. Christians deemed Hansen's disease as "unclean." Fear, panic, and misinformation drove despair and cruelty into people affected by Hansen's disease and resulted in brutal atrocities.

The most famous Hansen's disease colony is named Kalaupapa, situated on Molokai Island, Hawaii. Kalaupapa's scenery is gorgeous. Its soft, sandy beaches are adorned by vibrant flowers and lush, idyllic jungles. But the colony was still a prison. Thousands of Hansen's disease patients lived and died inside its thick, stone walls. As antibiotics for Hansen's disease became available, people slowly realized this method of quarantine was extremely unethical. Healthcare organizations today emphasize Hansen's disease patients don't need to be isolated.

The number of people living in Kalaupapa dwindled dramatically. In 2015, only six people lived there. They are all very old. Most were taken here when they were children. They remain because they know very little about their former world. Paradoxically, they even feel a strange sense of attachment to the colony. But what should the US do to Kalaupapa once its last resident dies? Should we open it up to tourists, or preserve it in peaceful solitude? Because of the sheer stigma these poor patients faced, there are myriad questions and debates shrouding the fate of Kalaupapa.

Roots of Stigma

At the dawn of the Middle Ages, people believed patients suffering from Hansen's disease were in purgatory—a place where Christians make up for their sins by suffering. To reduce their time in purgatory, Christians could care for or donate to Hansen's disease patients. Hansen's disease colonies were usually isolated from town, but patients could still beg for food or money, trade items, and/or pray for their helpers. A village consisted of a few cottages surrounding a chapel/church. The buildings were kept clean. Patients were fed a varied, healthy diet. Gardens of fragrant flowers and plants cheered dreary spirits. Patients even occasionally visited their family and friends.

But after the Black Death pandemic, things changed a lot. People lived in constant fear of diseases. Suddenly, the leprosy villages turned from hospital to prison. These colonies were completely cut off from society, and Hansen's disease patients suffered from extreme abuse and discrimination.

Hansen's Disease Around the World

In China, Hansen's disease colonies are isolated on islands and mountains. In each community, about 100 people live in mudbrick houses clustering a courtyard. Some are blind or have amputated limbs. Most people are cured of Hansen's disease but suffer from life-changing disabilities. People able to work cultivate small farms and tend to fish ponds. Some even own small tea shops and tailoring businesses. Games are played in courtyards, and schoolchildren occasionally come to sing for people living in these villages. Some people have access to electricity and own televisions. Hansen's disease colony populations have been declining since 1980, but still house a substantial number of people.

In the US, a Hansen's disease colony named Carville was established in 1894. At a first glance, Carville looked like a normal town. People played sports in a park. There were laundry services, a fire department, and even a bicycle repair shop. However, the colony started off as a poor, ramshackle smattering of "renovated" slave cabins in the swampy, malaria-filled marshes of Louisiana. Occupants were scorched in the summer and frozen solid in the winter. Marriage was banned. An on-site jail housed runaways from the village—a prison within a prison. People's identities were stripped away: residents were strongly advised to create new names for themselves to lessen the wicked stigma they faced.

To combat the evils of discrimination, a patient with the pseudonym Stanely Stein established a newspaper dubbed *The Star*. *The Star* featured articles written by Hansen's disease patients and caught the attention of people across the US. Despite the challenges Stein faced (he was completely blind), he was able to drastically improve the rights of people suffering from Hansen's disease with his poignant newspaper. Thanks to Stein's efforts, Carville received a makeover, especially after an antibiotic for Hansen's disease was synthesized in the 1940s. In 1946, residents were finally allowed to vote. The ban on marriage was lifted. Homes and a hospital were built. People cultivated flourishing gardens. Infrastructure improved. The Hansen's disease colony even hosted a small Mardi Gras festival. On the other hand, federal laws struggled to catch up with the improving villages. The discriminatory laws surrounding Hansen's disease were only removed decades later after science told the truth.

Fighting Back

Diagnosing Hansen's disease is easy in our high-tech world today. In a skin biopsy, infected skin and/or lesion samples are stained to reveal bacteria or scanned by PCR tests to sniff out bacterial DNA. Lepromin tests inject inactivated *M. leprae* under the skin. If a person's body reacts to the antigens (special bacterial proteins that induce an immune response) on the bacteria, he/she is probably infected. Lepromin tests could tell doctors if you contracted Hansen's disease recently or a long time ago. However, lepromin tests are usually too expensive and too slow to use for effective testing.

Treating Hansen's disease is also a breeze. Before an antibiotic was invented, patients were injected with chaulmoogra nut oil due to its alleged antimicrobial properties. Nobody knows if the oil worked, but some Hansen's disease antibiotics were derived from chaulmoogra nuts. In any case, chaulmoogra oil injections were agonizing and contained toxic cyanide. A proper antibiotic called promin was made in the 1940s. Later, a similar drug called dapsone was invented in Carville. Both medications worked well at first, but Hansen's disease eventually developed antimicrobial resistance. However, the disease was quickly clobbered again with multidrug therapy.

Down to Earth

Though toppled from its former glory, Hansen's disease still exists in the US. In 2023, doctors treated a fifty-four-year-old landscaper in Florida suffering from the disease, one among many victims of a Hansen's disease outbreak. After complaining to dermatologists for weeks about an itchy rash, a biopsy showed he had, in fact, contracted Hansen's. The culprit? Hansen's disease flourishes in the soil. Though the bacteria cannot survive on its own, it can infect soil-loving amoeba and wait for a larger, more chonky host to contract the disease. The landscaper, caked in dirt after a hard day's work, probably contracted the disease from these amoebas.

On a larger scale, soil also explains why armadillos are common hosts for Hansen's disease. Though it might not be obvious at first, armadillos are burrowing animals. Their immense burrows can stretch a gargantuan twenty-four feet wide and plunge five feet deep!

A Tragic History

Despite all our advances in medical science, Hansen's disease still squelches around in dirt worldwide. But perhaps more shameful is the filth spewed out by discrimination surrounding Hansen's disease. At least thirty countries around the world still haven't taken down their diabolical Hansen's disease laws. In these countries, filthy, run-down Hansen's disease colonies still house thousands of people. Worse, these colonies lack electricity and clean water. People often exile *themselves* to the colonies or get dragged there by a higher power due to the crushing psychological effects of the disease. Kids who've contracted Hansen's disease are teased and bullied at school and even discriminated against by their teachers. Young adults suffering from Hansen's disease are turned down from universities—legally! People with Hansen's disease can't even marry people who are disease-free. Though organizations are trying to cut down the stigma surrounding Hansen's disease, some of them are tainted with corruption, discrimination, and deception. A few particularly dastardly organizations lie about the annual number of Hansen's disease cases to say the disease is going down! This isn't right, nor any other form of discrimination. Hansen's disease proves how easy it is to get picked on for standing out from the crowd. It is not right to harass the sick and vulnerable. It is not right to attack someone just because they are different. This must stop, now.

CHAPTER 19

THE BACTERIOPHAGE

W<small>E HUMANS HAVE REALLY SCREWED UP. ABOUT A CENTURY</small> ago, we discovered the miracle drug penicillin concocted from the wondrous *Penicillium* mold. Soon, penicillin was just one of many antibiotics leading the charge of a cast of microbe-killers. We used to fear bacteria, cowering and shivering under their rule. But equipped with antibiotics, we broke free from the ruthless grip of bacteria and banished these ruthless, single-celled menaces to the fringes of society. People across the world used antibiotics as a wondrous cure-all; even the smallest coughs and stomach-aches disappeared with just a few nifty pills. But as the years passed, bacteria grew stronger and stronger with each generation until they were completely resistant to the biological superweapon we had made. Leveled up by natural selection, microbes fought back with vengeance. Doctors were overwhelmed. The panicked media dubbed these yoked microbes "superbugs." Today, the super-bug crisis has reached a boiling point: in 2019 alone, 1.27 million people across the globe perished at the hands of superbugs. The worst is yet to come. In a matter of decades, superbugs will kill more people than cancer!

Not All is Lost

Though the doom and gloom wrought by superbugs could fill page after page in this book, we humans can still fight back. A viral superhero called the phage might save the day, and we've given them a special mention in the *Pseudomonas* chapter. But first, let's have a recap. Phages come in all different shapes and sizes. Some resemble crystals and long pieces of string, but most look like giraffe-spiders with Slinky-shaped necks. The oversized "head" of the phage is called the capsid. It protects the fragile RNA inside the phage like a suit of armor. The sheath—the spring-like body of the phage—is attached to stick-like legs dubbed tail fibers.

Tail fibers are unique to phages. No other organism has them. These seemingly innocent little legs allow the phage to lock on and destroy targets. These tiny critters crawl in the thousands on you, your pet goldfish, and your leftover pizza, and they kill super-bugs, as well as other sorts of bacteria. Each phage is honed to hunt down only a couple of species of bacteria. These tiny spiders on stilts are completely harmless to us! Not that phages are gentle creatures; they kill forty percent of the bacteria in the oceans each day and infect trillions and trillions of bacteria per second! To put it lightly, it's a massacre when these micro missiles of death rain down from the sky.

To attack, a phage zeroes in on its target and locks on to the bacteria's receptors. These keyhole-shaped openings are normally used to receive signals from other bacteria. However, phages literally unlock the receptors with their tail fibers, allowing them to inject their RNA into the bacteria. The phage then hacks the bacteria's DNA. This forces the bacteria to manufacture phage parts and proteins, which piece together to form countless baby phages. When the bacteria are full of phages, they release an enzyme that explodes the bacteria, causing the young phages to fly out like fireworks. This method of phage reproduction is called the lytic cycle; a rather gruesome way to make new phages!

There's another, perhaps more efficient, way to whip up baby phages. It's called the lysogenic cycle. During the lysogenic cycle, a phage's DNA fuses with bacterial DNA, creating a secret "cheat code" called a prophage. The bacteria continue to divide and reproduce as if nothing had happened. However, the prophage tags along with the rest of the bacterial DNA as it reproduces. If the bacteria are exposed to UV radiation or certain chemicals, the prophage activates, the bacteria manufacture new phages, then the hapless cell bursts to let these little interlopers out. But there are some phages who bend the rules of replication: they sneak out of the bacteria before any external conditions kick in!

Some phages are restricted to either the lytic or lysogenic cycle. Others are "bilingual" and can use both. But how does a phage decide how to use the lytic or lysogenic cycle? Well, it comes down to a simple answer: numbers. Sometimes, many phages cram together and attack one bacterium. Though phages are many times smaller than the smallest bacterium, things can get crowded. If they're too many phages, these clever tricksters opt for the lysogenic cycle to prevent driving their host population to extinction. If the phage population is scanty and the bacteria population is high, phages prefer the lytic cycle.

Phages to the Rescue

The OP bacteria-killing properties of phages are well-documented in the medical field and have saved numerous lives. A man with his chest cavity full of the deadly bacteria *Pseudomonas aeruginosa* was as fit as a fiddle in a few weeks after a phage injection...after suffering for years in the hospital! But if bacteria evolved resistance to antibiotics, won't bacteria develop ways to fight phages off too? Yes. Phages and bacteria have been fighting an evolutionary arms race for billions of years. This is where antibiotics come in. In order to become resistant to phages, bacteria have one caveat: they have to give up their resistance to antibiotics! With a deadly duo of phages and antibiotics, bacteria can be put in a Catch-22. A phage cocktail (mixture of phages) also devastates bacteria. GMO phages are also an option!

Tom Patterson, an HIV researcher at UC San Diego, was also saved by phage therapy. After a vacation in Egypt, he fell dreadfully ill from the superbug *Acientobacter baumannii*. These lethal bacteria also killed troops in the Iraq War when it was blown into the wounds of soldiers. *A. baumannii* is an expert at stealing antibiotic resistance genes from other bacteria; it was invincible to every antibiotic doctors threw at it. As worst came to worst, his wife assembled a team of phage hunters as a last resort. They sifted through reeking manure, films of slime, and pungent sewage in a desperate search for phages that could kill the superbug. Though there were trillions of phage strains to analyze, his wife and the phage hunters finally found a potential candidate after days of desperate searching. These heroic scientists pulled all-nighters for days to find a phage, and their effort paid off: with phage injections, Patterson recovered in two days!

Patterson's story was a medical miracle. It should've shunted tons of money into phage research. But it didn't. While lab studies and last-resort hospital cases are promising, pharma companies are unwilling to pour billions of dollars into phage therapy. In the early 1900s, some US pharmacies prepared to create phage therapies. But phages were considered "retro" in just a couple of decades after they were discovered. The phages are surprisingly low-tech and finicky. Some so-called "preservatives" in phage containers deactivated and destroyed phages. Phages are also less easy to mass produce than antibiotics. Specific phages need to be isolated for specific patients. A massive library of thousands of phages needs to be kept to determine what phage is best for an ailing patient. If you want to see an example of a phage database yourself, go online to phagesdb.org. The US Food and Drug Administration also takes a long time—anywhere from one week to eight months—to confirm product safety. And to top it off, no one wants to go dredging through sewage to look for phages.

We could've avoided our phage dilemma if we put more research into phage therapy earlier. These little guys were first discovered by French-Canadian microbiologist Felix d'Herelle and British microbiologist Frederick Twort. However, people who experimented with phages did a bad job with experiment design and lacked crucial elements in their studies. These horrific experiments sparked tons of controversy: people even squabbled on whether phages were real. The creation of the electron microscope in 1931 silenced the debate for good. With this powerful invention, people gazed upon things they had never before seen—including phages. Phages went from being the Bigfoot or Nessie of microbiology to common knowledge by the 1940s.

Things Get Complicated

The former Soviet Union liked the idea of tiny, killer, bacteria-demolishers. During World War II, the US stopped using phages while Russia, Georgia, and Poland continued to use these wonder viruses in favor of antibiotics. But after the war, earlier collaborations between Russia and the US soon soured. Russia was Communist; everything Communist was feared in the US. People who used Russian research were locked in irons. Even prominent scientists were suspected of "Communist sympathies." Tracy Sonneborn was convicted when he experimented with paramecium. Even scientific breakthroughs attributed to the Soviet Union were pushed out the door and considered "beyond our border." Russia, which contributed the bulk of phage research, shut doors on the phage's availability in the US. But in eastern Europe, phages are still rad. Georgia once churned out two tons of phages per week to treat patients—amazing considering how weightless these viruses are! In these countries, you can still walk to a pharmacy and buy a vial of phages.

And then the bad boys, antibiotics, came to town. After penicillin was discovered by Alexander Flemming in 1928, phages were unheard of for the rest of the 1900s. These marvelous, Slinky-necked viruses were forgotten, kaput in the West. Because of limited antibiotics supplies in war-torn Europe, Europeans leaned heavily on phages for treatment in WWII. But in North America, antibiotics were the kings of the pharma industry and were (ab)used constantly. During a speech, Alexander Fleming himself warned not to overuse his miracle drug. Superbug renegades were popping up as he spoke! But it's hard for us not to overuse a new, trendy product on the market. Think about the last time you promoted a new shirt or new video game console on Facebook or Twitter and barraged it with hashtags. We lost respect for the microbial monsters that haunted us, as well as the weapons we used to hunt them.

And that's where we are today. Phages could save us. But let's hop back to Tom Patterson's ordeal. The bacteria that infected him were killed for good. But when bacteria samples taken from his body were attacked with phages, the bacteria soon became impervious to phages…again. A new phage had to be added to the phage cocktail. Some of Patterson's doctors even placed a fair bet the deluge of antibiotics—not the phages—helped Patterson recover!

Phage therapy sure is messy. The bacteria push back; the phages push back. The never-ending brawl between bacteria and phages is like an eternal seesaw battle at the playground. Like in a pinball game, we never know what'll happen next! It's very fortunate we won the first few rounds of microbial pinball and enforced some public confidence in these spindly viruses. By 2030, the phage market is expected to grow to $60.7 million.

Another patient cured by phages was a girl suffering from cystic fibrosis with the superbug *Mycobacterium abscessus* trapped in her lungs. Her illness progressed until her chances of survival dropped to one percent and her whole body was overwhelmed with bacteria. A few scientists saved the day by sifting through a phage database. They found three adorably-named phages to cure her infection: Muddy, BP, and ZoeJ. (Phages have some very interesting names. Some have even been named Yoshi, Zelda, and ShayShayLita). After suffering for months, the girl was released from the hospital in just nine days.

Attacking the Chatterboxes

Phages attack chatty bacteria the most. Bacteria use quorum sensing to talk to each other. Just like how some of us rant about the new Nintendo Switch or Xbox. In one study, phages eavesdropped on *Vibrio* bacteria. These chatterboxes communicate with the chemical DPO, and receive them with a protein called VqmA—basically their mailbox. Phages can detect DPO by installing a variant of VqmA on themselves! Many other phages might have protein mailboxes too!

A Glorious New World

But even if phages are constantly listening in on the weekly *Bacterial Times* to set up their next attack, we humans don't know which phages will work against certain bacteria unless we test 'em. This is a tedious, painstaking task. This is where new tech comes in. Doctors have identified a new gene in phages that make single-gene lysis proteins (Sgls), which are primed for destroying certain bacteria. An algorithm called **D**ual-**b**arcoded **S**hotgun **E**xpression Library **S**equencing (Dub-seq) reads the Sgls genes of phages, allowing people to scan new phages. Sgls themselves could be made and synthesized into new antibiotics. If faced with resistant bacteria, scientists can just turn back to the phages to develop a new Sgl. The possibilities for Dub-seq are endless!

Phages could soon start popping up in stores around you. Mom might give you grape-flavored "phage juice" to treat strep throat. In factories, phages could be sprayed onto food to kill bacteria during food preparation. In the military, phages could be carried along into the heat of battle, being poured onto potential bioweapons or nursing soldiers' wounds. Pharmacies all around you might sell "phagobiotics:" double vials where you gulp down antibiotics and phages at the same time. But it's still a pain to collect phages from the sewers. It's impossible for a pharmacy to hold all the phages in the world. Would phages be killed during the purification process? Would someone tamper with them? Will phages mutate to attack our microbiome too if given the chance?

And yet, there are new, awesome, gene-detecting technologies to personalize phages for each person's illness. There are easier ways to purify phages. After being ignored for so long, phages are being put into the light again, after decades of lurking in classified documents and political disputes. These little beasties can and will save us: tiny, micro spiders on stilts.

EPILOGUE

Congratulations! You've reached the end of this book. The final page. Now it's time for a word from your local microbial buddies. Before you go off scrambling to play on your Nintendo Switch or run outside to dribble a basketball, we hope you learned something new about us in this book. You've met our pals: the necklace-like *Streptococcus*, the fuzzy-headed *Aspergillus*, the biofilm-forming *Pseudomonas*, and pesky influenza. But it isn't our nasty side we want you to understand and fear, though we do want you to understand what microbes can do. After all, this is a biography of famous microbes. (Us!) It is the effect we have on history that we want you to understand: the good, the bad, and the ugly, not to mention how helpful we are for keeping the world tidy and in balance.

Consider how we impact human lives. (That's you!) We make medicine, help support your immune system, prevent obesity, and keep you happy and determined enough to grind for that math test looming large on the horizon. Even the "bad guys" contribute to human lives. Our chum *E. coli* can cause fatal diarrhea but supports microbiome health and helps nature with the decomposition process. Think rot is bad? Imagine having to step over the carcasses of dinosaurs and wooly mammoths mixed with Revolutionary War soldiers on your way to school. As the old saying goes, every cloud has a silver lining.

Give your microbiome a pat on the back. You can't shrink down to the size of us microbes to give us a present, so why not keep us happy by eating healthy foods? Why don't you take a break from the computer and go run a few laps around the block? Or just simply take a break? We microbes can feel when you are stressed and grouchy, and we get grumpy too! When we're beefy and happy, we also shove bad microbes out of the way. Another motivational note to stay healthy!

And of course, though this is optional, we would like you to choose your favorite microbe. Are you interested in how we have shaped society...or just how cutesy some of us look? Regardless, we recommend you keep reading, keep learning, and of course, stay curious!

Goodbye for now,
Your Neighborhood Microbes

SELECTED BIBLIOGRAPHY

Alcami, A. (2020). Was smallpox a widespread mild disease? Science, 369(6502), 376-377. https://doi.org/10.1126/science.abd1214

Angelov, A., & Liebl, W. (2006). Insights into extreme thermoacidophily based on genome analysis of Picrophilus torridus and other thermoacidophilic archaea. *Journal of Biotechnology, 126*(1), 3–10. https://doi.org/10.1016/j.jbiotec.2006.02.017

Ashworth, J. (2022). The world's largest bacteria are visible to the naked eye | Natural History Museum. https://www.nhm.ac.uk/discover/news/2022/june/worlds-largest-bacteria-are-visible-naked-eye.html

Avershina, E., Ravi, A., Storrø, O., Øien, T., Johnsen, R., & Rudi, K. (2015). Potential association of vacuum cleaning frequency with an altered gut microbiota in pregnant women and their 2-year-old children. *Microbiome, 3*(1). https://doi.org/10.1186/s40168-015-0125-2

Banuls, A. L. et al. (2015). Mycobacterium tuberculosis: ecology and evolution of a human bacterium. JMed Microbiol, 64(11), 1261-1269. https://doi.org/10.1099/jmm.0.000171

Barbier, M., & Wirth, T. (2016). The evolutionary history, demography, and spread of the Mycobacterium tuberculosis complex. Microbiol Spectr, 4(4). https://doi.org/10.1128/microbiolspec.TBTB2-0008-2016

Benedict, K., & Park, B. J. (2014). Invasive fungal infections after natural disasters. Emerg Infect Dis, 20(3), 349-355. https://doi.org/10.3201/eid2003.131230

Bhattarai, N., & T Stapleton, J. (2012). GB virus C: the good boy virus? *Trends Microbiol.* https://doi.org/10.1016/j.tim.2012.01.004

Butler, J. C., Shapiro, E. D., & Carlone, G. M. (1999). Pneumococcal vaccines: history, current status, and future directions. Am J Med, 107(1A), 69S-76S. https://doi.org/10.1016/s0002-9343(99)00105-9

Casadevall, A. (2012). Fungi and the rise of mammals. PLoS Pathog, 8(8), e1002808. https://doi.org/10.1371/journal.ppat.1002808

Centers for Disease Control and Prevention. (n.d.). CDC. https://www.cdc.gov/

Cleveland Clinic: Every Life Deserves World Class Care. (n.d.). Cleveland Clinic. https://my.clevelandclinic.org/

Damaso, C. R. (2018). Revisiting Jenner's mysteries, the role of the Beaugency lymph in the evolutionary path of ancient smallpox vaccines. Lancet Infect Dis, 18(2), e55-e63. https://doi.org/10.1016/S1473-3099(17)30445-0

De Smet, J. et al. (2017). Pseudomonas predators: understanding and exploiting phage-host interactions. Nat Rev Microbiol, 15(9), 517-530. https://doi.org/10.1038/nrmicro.2017.61

Durzynska, J., & Gozdzicka-Jozefiak, A. (2015). Viruses and cells intertwined since the dawn of evolution. Virol J, 12, 169. https://doi.org/10.1186/s12985-015-0400-7

Edwards, L. (2010). *Microbes survive a year and a half in space.* https://phys.org/news/2010-08-microbes-survive-year-space.html

Fisher, C. R et al. (2018). The spread and evolution of rabies virus: conquering new frontiers. Nat Rev Microbiol, 16(4), 241-255. https://doi.org/10.1038/nrmicro.2018.11

Forterre, P. et al. (2017). Plasmid vesicles mimicking virions. Nat Microbiol, 2(10), 1340-1341. https://doi.org/10.1038/s41564-017-0032-3

Grabenstein, J. D., & Klugman, K. P. (2012). A century of pneumococcal vaccination research in humans. Clin Microbiol Infect, 18 Suppl 5, 15-24. https://doi.org/10.1111/j.1469-0691.2012.03943.x

Guglielmini, J. et al. (2019). Diversification of giant and large eukaryotic dsDNA viruses predated the origin of modern eukaryotes. Proc Natl Acad Sci USA, 116(39), 19585-19592. https://doi.org/10.1073/pnas.1912006116

Hammer, T. J. et al. (2017). Caterpillars lack a resident gut microbiome. *Proceedings of the National Academy of Sciences, 114*(36), 9641–9646. https://doi.org/10.1073/pnas.1707186114

Health Library. (n.d.). Mount Sinai. https://www.mountsinai.org/health-library *Health topics*. (n.d.). https://www.who.int/health-topics/

Horcajada, J. P. et al. (2019). Epidemiology and treatment of multidrug-resistant and extensively drug-resistant Pseudomonas aeruginosa infections. Clin Microbiol Rev, 32(4). https://doi.org/10.1128/CMR.00031-19

Johnson, N. et al. (2014). Vampire bat rabies: ecology, epidemiology and control. Viruses, 6(5), 1911-1928. https://doi.org/10.3390/v6051911

Kletzin, A., & Adams, M. W. (1996). Tungsten in biological systems. *FEMS Microbiology Reviews, 18*(1), 5–63. https://doi.org/10.1111/j.1574-6976.1996.tb00226.x

Lange, C. et al. (2022). 100 years of Mycobacterium bovis bacille Calmette-Guerin. Lancet Infect Dis, 22(1), e2-e12. https://doi.org/10.1016/S1473-3099(21)00403-5

Lauck, M. et al. (2014). GB Virus C coinfections in west African Ebola patients. *Journal of Virology, 89*(4), 2425–2429. https://doi.org/10.1128/jvi.02752-14

Lehman, N. (2010). RNA in evolution. Wiley Interdiscip Rev RNA, 1(2), 202-213. https://doi.org/10.1002/wrna.37

Levin, P. A., & Angert, E. R. (2015). Small but mighty: cell size and bacteria. *Cold Spring Harbor Perspectives in Biology*, 7(7), a019216. https://doi.org/10.1101/cshperspect.a019216

McInerney, J. O. et al. (2014). The hybrid nature of the Eukaryota and a consilient view of life on Earth. Nat Rev Microbiol, 12(6), 449-455. https://doi.org/10.1038/nrmicro3271

Medical Diseases & Conditions - Mayo Clinic. (n.d.). Mayo Clinic. https://www.mayoclinic.org/diseases-conditions

Muhlemann, B. et al. (2020). Diverse variola virus (smallpox) strains were widespread in northern Europe in the Viking Age. Science, 369(6502). https://doi.org/10.1126/science.aaw8977

Nasir, A. et al. (2014). The distribution and impact of viral lineages in domains of life. Front Microbiol, 5, 194. https://doi.org/10.3389/fmicb.2014.00194

Nigro, O. D. et al. (2017). Viruses in the oceanic basement. mBio, 8(2). https://doi.org/10.1128/mBio.02129-16

Paez-Espino, D. et al. (2016). Uncovering Earth's virome. Nature, 536(7617), 425-430. https://doi.org/10.1038/nature19094

Panosian Dunavan, C. (2020). Superbug vs. superphage: A review of the perfect predator and an interview with Dr. Chip Schooley. Am J Trop Med Hyg, 102(1), 245-248. https://doi.org/10.4269/ajtmh.19-0770

Patterson, K. B., & Runge, T. (2002). Smallpox and the Native American. Am J Med Sci, 323(4), 216-222. https://doi.org/10.1097/00000441-200204000-00009

Reynolds, D., & Kollef, M. (2021). The epidemiology and pathogenesis and treatment of Pseudomonas aeruginosa infections: An Update. Drugs, 81(18), 2117-2131. https://doi.org/10.1007/s40265-021-01635-6

Schuster, R. (2022). Bacteria as big as tadpoles discovered in Caribbean swamp. *Haaretz.com*. https://www.haaretz.com/science-and-health/2022-06-22/ty-article/bacteria-as-big-as-tadpoles-discovere

d-in-caribbean-swamp/00000181-8bc8-db
5d-a9b3-eff893de0000

Shi, M. et al. (2018). The evolutionary history of vertebrate RNA viruses. Nature, 556(7700), 197-202. https://doi.org/10.1038/s41586-018-0012-7

Shi, M. et al. (2016). Divergent viruses discovered in arthropods and vertebrates revise the evolutionary history of the Flaviviridae and related viruses. J Virol, 90(2), 659-669. https://doi.org/10.1128/JVI.02036-15

Slifka, M. K., & Hanifin, J. M. (2004). Smallpox: the basics. Dermatol Clin, 22(3), 263-274, vi. https://doi.org/10.1016/j.det.2004.03.002

Steindler, L. et al. (2011). Energy starved Candidatus Pelagibacter ubique substitutes light-mediated ATP production for endogenous carbon respiration. PLoS ONE, 6(5), e19725. https://doi.org/10.1371/journal.pone.0019725

Thiele, K. (2019). The most ubiquitous and abundant species on Earth. Taxonomy Australia. https://www.taxonomyaustralia.org.au/post/the-most-ubiquitous-and-abundant-specie s-on-earth

Waters, E. (2003). The genome of Nanoarchaeum equitans: Insights into early archaeal evolution and derived parasitism. Proceedings of the National Academy of Sciences, 100(22), 12984–12988. https://doi.org/10.1073/pnas.1735403100

Wessner, D. R. (2010) Discovery of the giant mimivirus. Nature Education 3(9):61

Wolf, Y. I. et al. (2018). Origins and evolution of the global RNA virome. mBio, 9(6). https://doi.org/10.1128/mBio.02329-18

Woolhouse, M. E. et al. (2016). Assessing the epidemic potential of RNA and DNA viruses. Emerg Infect Dis, 22(12), 2037-2044. https://doi.org/10.3201/eid2212.160123

Yong, E. (2016). I contain multitudes: The microbes within us and a grander view of life. Random House.

Zhang, Y. Z. et al. (2018). The diversity, evolution and origins of vertebrate RNA viruses. Curr Opin Virol, 31, 9-16. https://doi.org/10.1016/j.coviro.2018.07.017

FURTHER READING

Barry, J. M. (2005). *The great influenza: The story of the deadliest pandemic in history.* Penguin Books.

Greger, M. (2020). *How to survive a pandemic.* Flatiron Books.

Mukherjee, S. (2023). *The song of the cell: An exploration of medicine and the new human.* Scribner.

Yong, E. (2016). *I contain multitudes: The microbes within us and a grander view of life.* Random House.

www.ingramcontent.com/pod-product-compliance
Lightning Source LLC
Chambersburg PA
CBHW071158210326
41597CB00016B/1592